Environmental Protection: Regulating for Results

Environmental Protection: Regulating for Results

Edited by

Kenneth Chilton
and Melinda Warren

Routledge
Taylor & Francis Group
LONDON AND NEW YORK

First published 1991 by Westview Press, Inc.

Published 2018 by Routledge
52 Vanderbilt Avenue, New York, NY 10017
2 Park Square, Milton Park, Abingdon, Oxon OX14 4RN

Routledge is an imprint of the Taylor & Francis Group, an informa business

A CIP catalog record for this book is available from the Library of Congress.
ISBN 13: 978-0-367-01242-7 (hbk)

Contents

Tables and Figures

Tables

Figures

Acknowledgments

The preparation of this book was supported by the Center for the Study of American Business (CSAB) at Washington University in St. Louis. The individual chapters reflect the Center's focus on research directed at environmental concerns in 1990.

Several of the chapters, those by Thomas DiLorenzo, Lester Lave, Margaret Maxey, and S. Fred Singer, resulted from CSAB's Key Issues Forum on Environmental Regulation. The Key Issues Forum was conceived to place special emphasis on the environment by engaging some of the leading thinkers in a variety of areas to prepare reports based on their current research. The remaining chapters are the products of the Center's own in-house research projects.

The co-editors would like to express special appreciation to the staff of the Center for the Study of American Business for their professional preparation of this book. The work of Rachel Kalin, who painstakingly edited and prepared many of the chapters, and the talented typing and layout skills of Mary-Eileen Rufkahr stand out. Producing camera-ready copy is truly an artform, requiring patience and skill. We also appreciate the keen insights offered by Center Director Dr. Murray Weidenbaum, especially his review of initial drafts of the research studies that led to this volume.

Last of all, we gratefully acknowledge the efforts of each of the chapter authors. Preparation of individual chapters required much coordination and goodwill to produce a homogeneous tone to the book. Our Key Issues Forum authors also gave thought-provoking talks at Washington University, based on their research. We hope "students" of environmental regulation, on-campus or off-campus, will benefit as greatly from reading these chapters as the faculty, students, and friends of Washington University did by hearing these talks in person.

Kenneth Chilton
Melinda Warren

Preface

Concern for the environment is reaching a new high in America. A variety of national polls confirms this assertion. An August 1990 survey by the Roper Organization showed that 70 percent of all Americans favor a major effort to improve the environment, up from 56 percent in 1987. A July 1989 *New York Times*/CBS News poll indicated that nearly 80 percent of U.S. citizens agree that "protecting the environment is so important that requirements and standards cannot be too high, and continuing environmental improvements must be made regardless of cost." Any reelection-conscious policymaker must take these sentiments seriously. Thus, legislative and regulatory initiatives are certain to abound in the decade of the 1990s.

But how should public policy be crafted to meet these legitimate concerns to protect the environment while simultaneously meeting other important objectives such as promoting economic growth, increased employment, and improved international competitiveness? Answers to this question are required at two levels: (1) analyses of specific ecological concerns and appropriate policy responses and (2) general principles that apply to a wide variety of environmental issues.

The following chapters address the need for environmental protection at both of these levels. The specific issues of municipal solid waste management, global warming, and atmospheric ozone (smog) are analyzed in Part One, "Key Concerns." Part Two, "Putting Environmental Risks in Perspective," broadens the scope of this analysis. Chapters by Dr. Lester Lave on "Setting Goals for Environmental Safety" and Dr. Margaret Maxey on "Managing Environmental Risks" focus on the fundamental environmental concern of most Americans: what are the health effects of pollution and what should government officials do to protect public health?

Part Three presents two key principles for cost-effective environmental protection. Professor Thomas DiLorenzo offers a convincing argument, based on evidence from Eastern Europe, that it is not capitalism, *per se*, that pollutes. He also emphasizes that capitalism creates the wealth needed to pay for pollution abatement. Dr. Murray Weidenbaum suggests that to get the most bang for the environmental buck, public policy should make maximum use of the power of the marketplace. Thus, the villain in the usual tale of environmental degradation—profit-making organizations—can be made the hero if public policy is crafted to regulate for results.

Environmental Protection: Regulating for Results grew out of a long-standing research emphasis on environmental regulation at the Center for the Study of American Business as well as a new program established to direct added research resources toward these concerns. The new program, dubbed the Key Issues Forum on Environmental Regulation, produced the chapters on global warming by S. Fred Singer, setting environmental safety goals by Lester Lave, Margaret Maxey's discussion of the environmental ethic, and Thomas DiLorenzo's answer to the question, "Does Capitalism Cause Pollution?"

This book presents a perspective on environmental regulation that is underreported in the national media. For those who maintain an open mind on this issue, the information presented in the following chapters will be educational and may fundamentally influence their views on how to best protect the environment. More than merely being concerned and apprehensive about pollution, the reader will be equipped with a results-oriented framework for analyzing current and future environmental issues.

Part One: Key Concerns

Municipal Solid Waste Management and Myths

America may be facing a crisis in municipal solid waste management, but it is not one caused by lack of solutions, writes Kenneth Chilton. Recycling, incinerator, and landfill technologies are up to the task, but the American public is its own worst enemy in resolving the problem of disposing of the discards from its abundant standard of living.

Fueled by environmental misinformation, Americans seem to be demanding zero-risk disposal of household wastes. Far too often, federal, state, and local officials respond with piecemeal political solutions that do little to address the fundamental issue of providing safe and cost-effective waste management.

This chapter seeks to answer a variety of key questions. What are the real problems and how might they be solved? What are the respective roles for recycling, for waste-to-energy plants, and for landfills? How might the power of the marketplace be harnessed to reduce the volume and toxicity of municipal solid waste and to make the siting of waste disposal facilities palatable? What should be the respective roles for local, state, and federal agencies in managing municipal wastes?

While the solutions to America's municipal waste problems are, by and large, technologically simple, they are by no means easy to implement. Reducing the level of environmental mythology and hysteria is a necessary first step down a long road to insure adequate capacity of safe and economical waste management for present and future needs.

Global Warming: Do We Know Enough to Act?

Concerns about the atmospheric greenhouse effect have spawned both doomsday scenarios and legislative proposals to stabilize the climate. Policies such as limiting the use of coal and putting a cap on carbon dioxide emissions are being considered to address global warming. For the most part these remedies would be ineffective and would exact a costly toll on U.S. economic growth. Moreover, these policies would impact mainly on the poorer segments of the population. If applied globally, such prescriptions would be denounced as a scheme to stop development in the third world.

Dr. Singer stresses the importance of examining the scientific facts before implementing policies that might prove counterproductive. Observed trends in global temperature readings do not agree with expectations from global warming theories. A large temperature increase of about 1 degree Fahrenheit occurred between 1880 and 1940, well before human influences were important. (Despite the growth of heavy industry during that period, the amount of fossil fuels burned for energy was small compared with that burned today.) A temperature decline occurred between 1940 and 1965, followed by a sudden warming of about 0.3 degrees Fahrenheit since 1975—too short a period to discern a trend.

What should concern policymakers most is a very rapid change in climate, one to which the U.S. economy cannot adjust, writes Dr. Singer. In fact, even the most drastic measures to limit fossil-fuel burning can do little to stave off a rise in the atmospheric greenhouse gases—delaying a doubling from perhaps the year 2040 to only 2045. Furthermore, a crash program to resolve carbon dioxide emissions could cost the U.S. economy a trillion dollars, according to some estimates.

Even if significant warming were to occur in the next century, the available evidence indicates that the net impact may well be beneficial. Yale economist William Nordhaus has pointed out that ". . . those who argue for strong measures to slow greenhouse warming have reached their conclusion without any discernible analysis of the costs and benefits. . . ."

Energy conservation, efficiency increases, and use of non-fossil fuels are all prudent policies, as long as they are cost-effective, says Dr. Singer. But more drastic, precipitous—and especially unilateral—steps to delay the putative greenhouse impacts can cost jobs and reduce economic prosperity without being effective.

Battling Smog

Nearly half of the 83 areas not meeting the nation's air quality standard for atmospheric ozone (smog) in 1989 likely will still be nonattainment areas in the year 2000. But what does this projection mean to citizens in these areas and what should be done?

This analysis by Anne Sholtz-Vogt and Kenneth Chilton addresses such questions about urban smog as: (1) what are the health and other effects of ozone? (2) what causes ozone buildup? (3) what are the costs and benefits of reducing ozone levels? (4) how will the new Clean Air Act amendments alter the attack?

Sholtz-Vogt and Chilton cite extensive medical research collected by the Environmental Protection Agency (EPA) that shows ozone's short-term effects, on average, to be relatively mild—even at levels as high as 50 percent above the air quality standard. These effects are also reversible: most healthy adults typically recover in less than six hours from acute effects caused by a two-hour exposure to ozone levels as high as three times the standard.

This chapter also analyzes government data to estimate benefits and costs for programs designed to reduce ozone. The authors find:

> The best estimates of the costs and health benefits of reducing VOC emissions (a key ingredient in ozone formation) in nonattainment areas indicate a ratio of health benefits to abatement costs ranging from 0.2 to 0.4 for a 30 to 40 percent reduction. When costs exceed benefits, that is a warning signal that the resources of the nation are being used inefficiently.

The authors contend that the Clean Air Act reauthorization passed in 1990 does not consider adequately the fundamental changes needed to assure that public health is protected with less economic disruption. They offer four fundamental reforms to accomplish this more balanced objective:

- Require the EPA administrator to consider cost-effectiveness when promulgating Clean Air Act regulations.
- Change the definition of nonattainment to be more consistent with the nature of ozone pollution and of the public health risk it poses.
- Establish primary air quality standards to protect the public against *unreasonable risk* of *significant* adverse health effects. (The current requirement to set standards to provide an *adequate margin of safety* against *any* possible health consequences is tantamount to legislating "zero risk" from air pollutants.)
- Reduce the number of specific control measures required in the Clean Air Act. Encourage states to use innovative approaches that fit local circumstances.

Part Two: Putting Environmental Risks in Perspective

Setting Goals for Environmental Safety

Public policymakers have a legitimate concern with carcinogenic pollutants in the environment. Nonetheless, writes Professor Lester Lave, frequently too many resources are focused on relatively remote risks where public outcries are greatest—EDB, Alar, and so on.

Dr. Lave suggests that regulators determine where federal dollars will do the most good using the tool of cost-effectiveness analysis. Various sources of cancer risks cost different amounts to reduce. For example, protecting citizens from radon in buildings would cost much less per cancer prevented than would eliminating all traces of benzene in drinking water. Lave suggests that regulators start by addressing those situations which pose a significant risk—those producing an observable effect—and which require fewer dollars to alleviate.

Manmade chemicals are not really major sources of cancer, as most people assume. A study for the National Cancer Institute showed that food additives and industrial products, combined, are responsible for less than six percent of all cancers, while almost two-thirds of cancers could be prevented through changes in diet and smoking habits. In fact, the Institute's plan to cut cancer deaths in half by the year 2000 does not include reducing carcinogenic chemicals in the environment as an objective.

In addition, Lave notes the inadequacy of current federal policy on carcinogens. Most regulations are based on the "one-hit theory," which says that exposure to even one molecule of a carcinogenic substance can trigger the development of cancer. The problem is that scientists can detect smaller and smaller amounts of chemicals. To ban all substances containing any amount of a known carcinogen has become a ludicrous proposition.

Dr. Lave believes that policymakers should expend resources where they will have the most impact. To determine which areas these are, regulators must employ the tools of risk analysis and cost-effectiveness analysis.

Managing Environmental Risks: What Difference Does Ethics Make?

The theme of Dr. Margaret Maxey's analysis is the need to make tough environmental choices. She examines how Americans have come to embrace a new "environmental ethic" to replace the ethic of utility—the greatest good for the greatest number.

Dr. Maxey notes that as technology has become more sophisticated, Americans have become more fearful of it. The assumption is that Nature is noncarcinogenic and that industrial society is the source of all cancers.

Maxey asserts that ignoring natural risks and focusing only on manmade risks is illogical. As an example of regulatory inconsistency, she cites the case of polynuclear aromatic hydrocarbons (PAHs), which have been regulated at 0.03 micrograms per liter in drinking water. One PAH, benzo(a)pyrene, occurs in detectable levels in charbroiled sirloin steak, bread, and lettuce. If the EPA regulated these foods at the same risk level as it regulates water, we would only be permitted to eat a 10-ounce steak every two months or 2 slices of bread daily to the exclusion of all other PAH-containing substances.

She concludes:

> In the real world where priorities must be set and public revenues effectively allocated, regulatory officials are to be commended rather than condemned for comparing the costs of alternative methods of reducing *actual* harm. Merely outlawing hypothetical risks takes little courage and produces only hypothetical benefits.

Part Three: Markets and the Environment

Does Capitalism Cause Pollution?

Corporations are often accused of despoiling the environment in their quest for profit. Free enterprise is supposedly incompatible with environmental preservation, thus requiring government regulation to keep it under control. According to Professor Thomas DiLorenzo, this thinking is the basis for proposals to greatly expand environmental regulation, including a proposal to give the EPA cabinet-level status.

If the profit motive is the primary cause of pollution, however, why is there so much pollution in socialist countries like the USSR, China, and in Eastern Europe? Indeed, the socialist world suffers from the worst pollution on earth. In Czechoslovakia, buildings are eroded from air pollution and hills are bare because some fields are toxic to the depth of one foot. In Poland, a quarter of the soil may be too contaminated for safe farming. In fact, in four of the six East European countries, a quarter to a third of the forests show signs of dying from air pollution.

These examples of environmental degradation suggest some valuable lessons. First, it is not free enterprise *per se* that causes environmental harm, says Dr. DiLorenzo. The heart of the problem lies with the failure of public institutions, not the free enterprise system.

Plundering of the environment in the socialist world is an example of what biologist Garrett Hardin called the "tragedy of the commons." The inclination of each individual under communal property ownership is to abuse or deplete the resource before someone else does.

The two pillars of free enterprise—sound liability laws that hold people responsible for their actions and the enforcement of private property rights—are important stepping stones to environmental protection, maintains Professor DiLorenzo.

Making the Marketplace Work for the Environment

Dr. Murray Weidenbaum focuses attention on the use of economic incentives to manage environmental problems. Opinion surveys show that a substantial majority of American citizens want to improve the environment "regardless of cost," but they do not want to lose American jobs in the process. Economic incentives can help reduce environmental pollution in a manner that is less disruptive to economic growth.

One area where market incentives are needed is in the establishment of hazardous waste sites, writes Dr. Weidenbaum. Because of strong opposition on the part of the citizens living near potential sites, not a single major new disposal facility has been established in the United States since 1980. The objections of local residents are very logical—society as a whole benefits while local citizens pay the costs of possible leakage and depressed property values. On the other hand, resistance to hazardous waste facilities imposes significant costs on society as a whole.

According to Professor Weidenbaum, the solution to siting new, safer facilities lies in using economic incentives. If citizens know that when they accept a disposal facility they will get something tangible in return—a new school building or firehouse they could not otherwise afford—opposition will decrease. In one community, an entrepreneur's vague promise that the facility would provide $1 million a year in tax revenues to the local government was not enough for the citizens to accept an incinerator. But when he offered to pay property taxes for all the town's landowners over the next twenty-five years, a majority of the citizens voted to accept the proposal.

Economic incentives can also encourage the prevention of pollution, writes Weidenbaum. Pollution fees or taxes on the amount of pollutants discharged would motivate manufacturers to change their production processes to reduce the amount of wastes created or to recycle them. Pollution taxes are not a system of punishment but a way to increase the costs of polluting, says Dr. Weidenbaum. Business people understand the idea of minimizing costs to maximize profit.

Professor Weidenbaum concludes that the reform of environmental regulation is ultimately a consumer issue. It is consumers (citizens) who benefit from a cleaner environment and who also bear the burden of compliance through higher prices for goods and services. Americans must face the fact that cleaning the environment is as much an economic issue as a moral one.

PART ONE

Key Concerns

1

Municipal Solid Waste Management and Myths

Kenneth Chilton

Trash—technically, municipal solid waste—is on the minds of nearly all Americans today. Most people are disturbed by televised reports of barges of garbage from New York on a worldwide search for a port to call home, medical wastes floating up on East Coast beaches, and tractor trailers that carry produce from the midwest to the east and return laden with garbage.

The public's concerns about waste disposal deserve a serious response. The problem of handling the nation's garbage is real and increasingly difficult to solve. Almost every important public issue, however, generates emotional responses. Within limits, this is helpful. Dramatic headlines arouse public support for change. But experience also tells us that there are important limits to the good that public outrage can do.

Workable solutions to complex issues do not arise from ten-second sound bites. Effective answers require a different approach—the less dramatic but equally essential step of careful analysis. To address the problem of properly managing America's municipal solid wastes, we must put our trash woes in proper perspective and match public policy to the critical components of the dilemma. In the process, it is necessary that several widely held myths be demolished and attention focused on the substantial issues, ones that cannot be placed readily on a bumper sticker.

Fighting Mythology

Myth #1: America Is Running Out of Landfill Space

What are some of the myths that are widely accepted about America's trash-disposal problem? First of all, most people believe this situation is a "crisis" because, supposedly, we are running out of safe landfill space. After all, 50 percent of the landfills in the United States will close in the next five years; 80 percent will close in twenty years.[1] The specter of a significant portion of the 190 million tons of consumer trash expected to be disposed of in the year 2000 having no place to go is truly worrisome.

Environmental Protection Agency (EPA) surveys indeed do show landfill capacity declining dramatically over the next decade. There were 5,500 operating landfills processing 187 million tons of consumer and industrial solid wastes in 1988. Less than 2,200 of these sites will remain open in the year 2000, and they will be capable of handling only 76 million tons of trash a year.[2]

In addition, recent actions by state and local governments will accelerate the closure of substandard landfills. This is both good news and bad news: future landfilling practices will be more environmentally sound, but the cost of waste disposal will be substantially higher and a solid waste disposal capacity crunch will be that much more likely. Proposed EPA landfill guidelines will likely accelerate closings. These guidelines exempt landfill owners from costly closure and cleanup requirements if their facilities are shut down within one-and-a-half years of the adoption of the guidelines.[3]

However, the fact that landfills close is not news. By design, most landfills remain open only 10 or 20 years. The landfill dilemma is caused by the paucity of new landfills. Current construction rates will add only about four million tons of new capacity a year. Figure 1.1 reflects the consequences of this greatly reduced rate of landfill siting. Depending upon the amount of recycling and combustion (incineration) being done, a capacity crisis could be expected as early as 1997 or as late as 2000.

Building safe landfills hardly requires space-age technology. The key requirements are systems to collect and process leachate—the liquid product of moisture seeping into the site

FIGURE 1.1
The Impending Landfill Crunch
(landfill capacity in millions of tons per year)

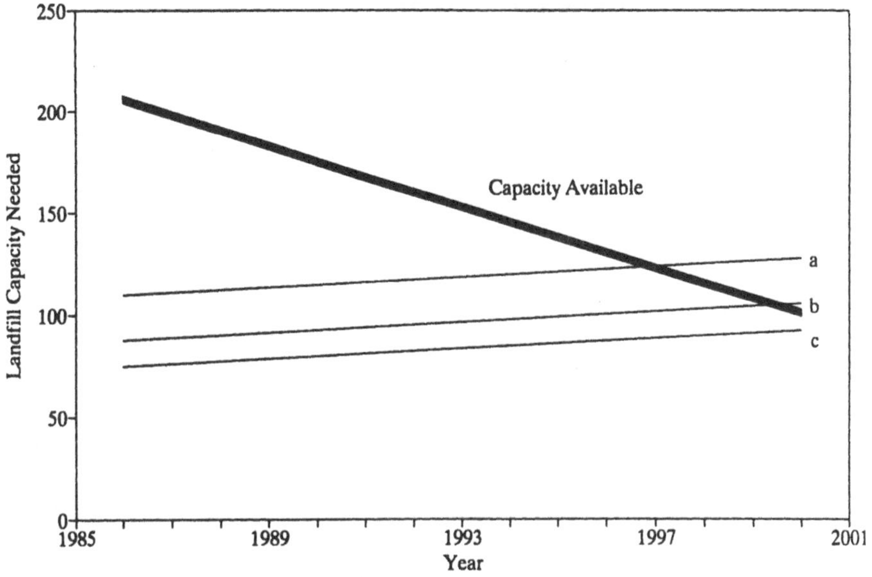

a 15% of waste stream recycled; 15% of waste stream burned in resource recovery plants.
b 25% recycled; 25% resource recovery.
c 25% recycled; 35% resource recovery.

Source: *Landfill Capacity in the Year 2000* (Washington, D.C.: National Solid Wastes Management Association, 1989), p. 4. From 1986 EPA data.

combined with organic and chemical components stored there—and a method to collect and vent (or burn) methane gas generated by organic decomposition. Modern sites have a single- or double-composite liner or a liner of clay, a collection system, and pumps to send leachate to the surface for processing. They also have a system for monitoring and safely disposing of methane gas, often burning the gas to generate heat to be used by a nearby plant or public building.

Of course, not all locations are environmentally acceptable. A high water table or a porous subsurface would present serious leachate and groundwater-contamination problems. Long Island is not a desirable site for a landfill, for example, because of its high water table.

Yet, we are not running out of sites that are geologically suited for landfills. A recent environmental survey of less than half of the state of New York pinpointed potential sites totalling 200 square miles. In general, geologically acceptable sites abound in America.[4]

Landfills, even when made environmentally safe, are difficult to make desirable to the surrounding community. Thus, the siting of new facilities presents what economists term a public goods, or free-rider, problem. A large geographic area benefits from concentrating trash disposal at a relatively small site. However, the landfill lowers neighboring property values and causes affected citizens to oppose the facility. The observed Not-In-My-Backyard (NIMBY) syndrome is not difficult to understand. Moreover, NIMBY naturally leads to LULU—charges that siting the landfill in the targeted area is "Locally Unacceptable Land Use." It is NIMBY that is the true impediment to siting, not a lack of technological expertise or safe geological sites.

Although the public goods problem is not new, the political potency of NIMBY and LULU is on the rise. In the early 1970s, municipal landfills were constructed at a rate of 300 to 400 a year. In the decade of the 1980s, this rate decreased to between 50 and 200 a year.[5] Heightened environmental awareness has increased resistance to new landfills even while the environmental risk from these facilities has decreased significantly.

Myth #2: Americans Are Trash Junkies

Americans are admonished almost daily about their inordinate wastefulness. A typical U.S. citizen discards 3.5 pounds of garbage a day (1,300 pounds a year). U.S. consumers toss 1.6 billion pens, 2 billion razors and blades, and 16 billion disposable diapers into the trash every year.[6]

We are told that this wastefulness is far greater than our foreign counterparts and is a symptom of the "throw-away society" that ease-loving Americas have created. For example, an EPA report on solid waste notes that "an American generates approximately one pound per day more waste than his/her counterpart in West Germany."[7] The Japanese recycle 40 percent of their municipal solid waste compared to our puny recycling rate of 13 percent.

What we are not told, however, is that municipal solid waste (MSW) data—foreign information and ours—are of uncertain quality. Most U.S. figures come from estimates of "materials flow" generated by a computer model written by Franklin Associates for the EPA. The EPA/Franklin model calculates the composition of MSW by weight, in terms of different materials and products in much the same way that economic activity can be broken down into industrial categories using input/output tables.[8] The EPA/Franklin model estimates an average of 3.5 pounds of daily MSW per person in the United States. However, data from several landfill excavations indicate daily disposal rates of less than 3 pounds per person.[9]

Furthermore, the data quoted for daily solid waste disposal in other nations most often exclude recycled materials. When "net discards" are considered, the U.S. figure (according to the EPA/Franklin model) is about 3.2 pounds, for West Germany 2.6 pounds, Sweden 2.4, Switzerland 2.2 and Japan, a surprising 3.0 pounds. Australia, Bulgaria, Canada, Hungary, New Zealand, and Korea have generation rates similar to the United States.[10]

Moreover, our "throw-away society" is not necessarily more wasteful than less-developed societies. According to Professor William Rathje, an archaeologist at Arizona State University who specializes in excavating landfills, "The average household in Mexico City produces one-third more garbage a day than does the average American household."[11]

Myth #3: Plastics Are the Problem

The leading villain in the municipal solid waste saga is the *unnatural* substance known as plastic. Plastic products shrink our landfill capacity by stubbornly refusing to biodegrade. According to some environmentalists, they also threaten our groundwater supplies and our ozone layer.

Plastic is ubiquitous. It is in disposable diapers and consumer packaging—beverage containers, microwave dinners, and shrink packs for Teenage Mutant Ninja Turtle toys. Worst of all, say the environmentalists, are polystyrene clam shells and beverage cups that are so popular with fast food outlets. Even the grocery stores have dared to replace the familiar brown grocery bag with thin, odd-looking plastic sacks.

One of the most interesting findings of Professor William Rathje and his "Garbage Project" is the lack of biodegradation in modern landfills. Perfectly preserved 30-year-old newspapers and ears of corn are typical of landfills in any climate. Each day's trash is covered with a layer of dirt to reduce odors, prevent runoff and discourage animal (and human) scavenging. As a result, biodegradation and photodegradation, which require sunlight, moisture and/or air, do not occur within the landfill except for yard wastes and other organic materials which degrade at a very slow rate.

Trying to make packaging "biodegradable" might have some benefits for reducing litter but may well be counterproductive for making landfills more environmentally sound. Biodegradable plastic often requires that more plastic be used to compensate for the loss in strength resulting from the use of a degeneration agent, such as cornstarch. Further, biodegradable plastics do not liquify. Totally degraded plastic is still plastic. The plastic has not disappeared, it has simply gone to pieces.

The fact that plastic does not biodegrade is not a sin; it is a virtue. Because it is inert, plastic does not release chemicals into streams or groundwater.

The grocery store that offers plastic bags instead of paper bags typically does so because the bags take less storage room. Given the lack of natural decomposition of paper in a landfill, it seems logical that plastic grocery sacks also take up less landfill space. Once again popular wisdom does not square with reality.

Furthermore, polystyrene—public enemy number one in the environmentalist's view—is not the threat that it is supposed to be. Its impact on stratospheric ozone depletion comes from the use of chlorofluorocarbons (CFCs) as blowing agents in the production process, not from disposal of fast-food clam shell containers and coffee cups. Using CFCs as a blowing agent for production of polystyrene is being discontinued as a result of the Montreal Protocol, an international agreement to phase out the production of CFCs. As for its role in wasting landfill capacity, studies indicate that all types of fast-food packaging combined comprise only 0.3 percent (by both weight and volume) of materials excavated.[12]

This information has not stopped local governments from banning the use of polystyrene by fast food outlets, however. In fact, "styro cops" now roam Seattle and other cities looking for fast food scofflaws to bring to justice.

Myth #4: Landfilling and Incinerating of Household Trash Must Stop

Another accepted axiom of the municipal solid waste crisis is that current waste management practices—landfilling nearly 73 percent of our trash, incinerating another 14 percent and recycling a paltry 13 percent—are unacceptable. Waste reduction is seen as the ultimate solution, decreasing the volume and toxicity of products and packaging prior to distribution or enhancing their recyclability. Recycling rates must be much greater, regardless of the current economics of recycling, say the environmentalists.

Landfills and incinerators are considered too harmful to human health and to the environment to be relied upon to any great extent. Landfill leachate contaminates drinking water. Methane gas buildup from decaying wastes poses a threat to life and limb as well as contributing to smog. Incinerators produce undesirable air pollutants, especially dioxins, and ash laced with heavy metals such as lead and cadmium poses a threat to groundwater supplies.

Because of the widespread negative publicity given landfills and incinerators, the public may now believe that there is some superior "silver bullet" available to deal with municipal solid waste. A March 1990 public opinion survey conducted for the National Solid Wastes Management Association (NSWMA) indicates that 36 percent of those surveyed would not object to building new landfills in their communities, while 59 percent were opposed. Waste-to-energy plants fared somewhat better with 55 percent of respondents indicating that they would not object to such a facility in their community, while 37 percent would object.

An overwhelming majority, 92 percent, of American adults believe that "a major commitment to recycling will substantially reduce the nation's solid waste," according to NWSMA. Nearly half of the survey respondents say that the United States can recycle more than 40 percent of its garbage.[13] Nonetheless, even a huge increase in recycling, say to 50 percent by the year 2000, would leave nearly 95 million tons of garbage to be incinerated or disposed of in landfills. This fact leads to the next topic: "What are the alternatives for dealing with municipal solid waste?"

Reducing and Managing Municipal Solid Waste

There are four basic ways to reduce or manage municipal solid waste: source reduction, recycling, incineration, and landfilling. Source reduction takes place during the design and manufacture of products and packaging to minimize toxic content, reduce the volume of material, or increase the useful life (or recyclability) of products or packaging. Source reduction might be thought of as preventive medicine.

The actual disposal of municipal solid waste falls into one of the three categories shown in Table 1.1. This table indicates the percentages of solid waste disposed of in 1988 by recycling, incineration, and landfilling. It also shows EPA's stated goals for altering the role played by each of these waste management methods by 1992.

To reduce the amount of MSW going to landfills, the EPA has set a goal of increasing recycling rates from 13 percent to 25 percent. Incineration rates should also be increased—from 14 percent to 20 percent. These are overall goals, of course, and some regions of the country will need to rely more heavily on recycling and incineration than others.

Integrating Management Methods

In February 1989, the Municipal Solid Waste Task Force published its final report for the EPA, *The Solid Waste Dilemma: An Agenda for Action.* The Task Force emphasized an "integrated management hierarchy" based on the four fundamental methods of waste reduction and management outlined above. The report stressed that all four elements should be "integrated" into a system designed "to emphasize certain management practices, consistent with the community's demography and waste stream characteristics. . . . In an integrated waste management system, each component is designed so it complements, rather than competes with, the other components in the system."[14]

By choosing to refer to waste reduction and disposal as an "integrated management *hierarchy*," however, the MSW Task Force created the impression that source reduction is preferred to recycling, which is preferred to combustion (incineration), and last of all, landfilling. Many environmentalists and public policy-

TABLE 1.1
Methods of Municipal Solid Waste Disposal
(as a percentage of waste stream)

Management Method	1988	EPA Goal for 1992
Recycling	13%	25%
Incineration	14%	20%
Landfilling	73%	55%

***Source*:** Derived from EPA Municipal Solid Waste Task Force, *The Solid Waste Dilemma: An Agenda for Action* (Washington, D.C.: U.S. Environmental Protection Agency, 1989) and *Characterization of Municipal Solid Waste in the United States: 1990 Update* (Washington, D.C.: U.S. Environmental Protection Agency, June 1990).

makers seem to favor the notion of a waste management hierarchy over that of an integrated system, determined by local needs and resources. But, in a world of finite resources, we must use our resources in the most cost-effective way, rather than according to a hierarchical scheme that could result in high costs for small benefits.

Source Reduction. One method of source reduction is to ban the use of specific substances that may present significant risk in the disposal process. This is an approach that should be used sparingly, however, and with careful analysis both of costs and benefits.

Unintended side effects often accompany bans. For example, a Senate solid waste bill would prohibit the nonessential use of cadmium, including in pigments. The primary target seems to be the cadmium used in rechargeable batteries that can break down when incinerated into an easily dispersed and soluble toxic substance. But artists are outraged because there is no acceptable substitute for cadmium in producing bright red, yellow and orange hues.[15]

By and large, economic incentives are necessary to further reduce the volume and toxicity or increase recyclability and the useful life of products and packaging. One of the prime impediments to greater source reduction is that producers, for the most part, are not faced with disposal costs for their products or packaging. Auto batteries, tires, household batteries, and major appliances can present special problems for disposal facilities. Proposals to give manufacturers direct responsibility for disposal of these products have some merit but need to be flexible to allow for cost-effective solutions not foreseen when legislation is drafted.

Specific schemes that promote source reduction are no panacea. This approach is likely to be most appropriate for dealing with toxicity concerns. However, volume reduction can be better promoted by practices which bring home to consumers the true costs of waste disposal. This means increasing the price of some products to cover the expense of getting rid of them when they are no longer usable or needed.

Recycling. Recycling has nearly reached the level of public support given motherhood and apple pie. Recall that the National Solid Wastes Management Association survey found that 92 percent of American adults believe that a major commitment to recycling will substantially reduce the solid waste stream.

Figure 1.2 depicts estimates of the degree of recycling being done in the United States currently. The figure provides two sets of estimates, those calculated for the EPA by Franklin Associates and those reported by industry. Not surprisingly, industry estimates of recycling are higher than EPA figures.

Of course, the effect of recycling rates on the municipal solid waste disposal problem depends greatly on the portion of these materials in the total solid waste stream. Aluminum is recycled at a rate somewhere between 25 and 45 percent but constitutes only 1 percent of MSW. Paper and paperboard are recycled at a lower rate—between 22 and 28 percent—but they account for 41 percent of municipal waste.[16] Perhaps the greatest potential for beneficial recycling is residential yard wastes, which make up nearly 20 percent of MSW.

Incineration (Waste-to-Energy). Incineration has an important role to play in an integrated waste management system, particularly in those areas of the country where siting new landfills is especially difficult. Incineration reduces the amount of solid waste by 75-90 percent. Most existing or planned U.S. combustors are either "mass burn" systems that burn mixed, unprocessed MSW or "refuse-derived fuel" systems, which first mechanically process solid waste to produce a more homogeneous fuel.

FIGURE 1.2
U.S. Recycling Rates:
EPA and Industry Estimates

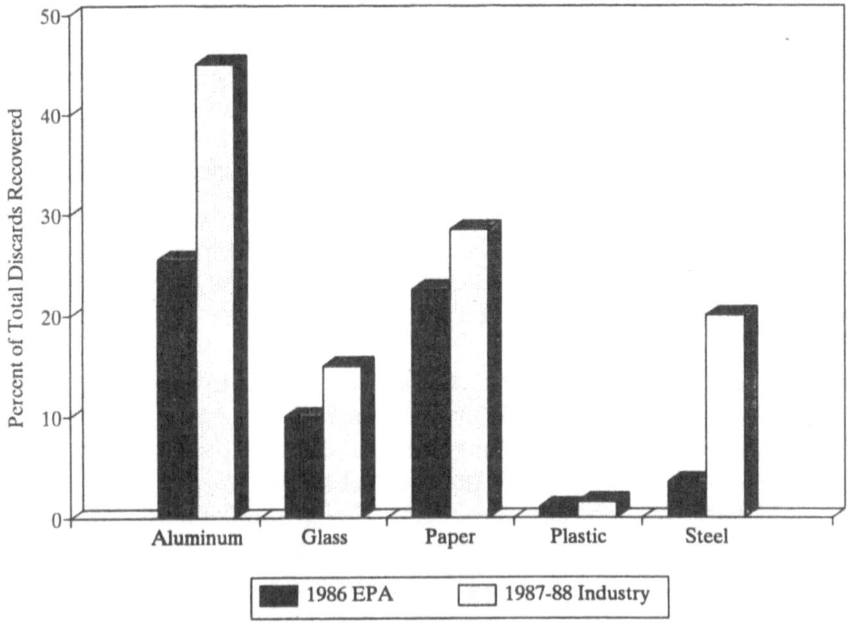

Source: Office of Technology Assessment chart from *Facing America's Trash: What Next for Municipal Solid Waste?* (Washington, D.C.: Office of Technology Assessment, 1989), p. 28.

As mentioned previously, public opinion surveys indicate that the majority of Americans would not object to a new waste-to-energy plant in their community. This favorable attitude is fairly recent, however, and may be somewhat affected by individual interpretation of the terms "object to" and "community." A February 1989 survey conducted by NSWMA asked individuals, "Would you favor or oppose placing a new waste-to-energy plant in your community?" At that time, 48 percent said they would *oppose* a new combustor, versus 39 percent who said they would *favor* one being sited in their community.[17]

Public opposition to incinerators results primarily from concerns about the presence of unhealthful residues of metals and organic chemicals in air emissions and in ash. The Municipal Solid Waste Task Force concluded, "Although combustion is not risk-free, a state-of-the-art combustor that is well operated should not present a significant risk to human health and the environment."[18] Nonetheless, uncertainties created by public opposition and the real possibility of "moving target" regulations on air emissions and ash disposal are slowing the pace of investment by cities and private firms in incinerators.

Interestingly enough, while we read frequently of higher recycling rates in foreign nations than in the United States, we hear far less about their greater reliance on incineration to manage their MSW. The Office of Technology Assessment's study, *Facing America's Trash*, reports the following facts about Japanese and European use of combustors:

- In Japan, over 1,900 cities (almost two-thirds of all municipalities) have incinerators of some sort. About one-fifth of all Japanese combustors are mass burn. Japan incinerates nearly two-thirds of its post-recycling MSW.
- The majority of European facilities are mass burn, waste-to-energy facilities. Sweden and Denmark combust more than half of their post-recycling solid waste and West Germany about one-third of its MSW after recycling.[19]

Plans for new incinerators in the United States should increase the portion of MSW disposed of in these facilities to be about 25 percent by the end of the century. Currently, there are more than 160 combustors (about 120 are waste-to-energy facilities; the remainder merely incinerate the trash without producing energy) that process from 10 to 15 percent of the municipal waste. About 45 facilities are under construction.[20] A fourth of the MSW entering incinerators must, in turn, be disposed of in landfills or otherwise stabilized. The net reduction of MSW in the year 2000 from this method is likely to be about 20 percent. This is the rate of MSW incineration that EPA hoped would be reached by 1992.

In spite of the appeal of disposing of garbage and simultaneously generating electricity, the economics of waste-to-energy plants is less compelling. Construction outlays can run as high as $500 million, although most seem to cost $350-450 million. The Public Utility Regulatory Policies Act requires public utilities to purchase electricity from such facilities at a price equal to the "avoided cost"—the cost for the utility to generate this energy.

Because waste-to-energy plants' costs are higher than the price paid by the public utilities, incinerators are not cost-effective without considering the foregone costs of alternative MSW disposal methods.

Waste-to-energy facilities, particularly, require an integrated approach to MSW management. Too much recycling can reduce the amount of "fuel" coming to the combustor. On the other hand, too little recycling—particularly, failure to handle yard wastes, glass and metals separately—can cause higher levels of nitrogen oxide emissions and variable temperatures that can interfere with clean burning. In addition, source reduction of certain plastics that may produce dioxins when burned, and recycling of household batteries and other products that may produce undesirable residues in ash, can enhance the safety of this waste management method.

But rather than take such an integrated view of MSW, many cities are relying on greatly increased recycling rates to head off a crisis. For example, Seattle decided to increase its recycling rate to 60 percent by 1998 rather than site new landfills or build an incinerator.[21] The city currently recycles at a 37 percent rate, the highest recycling rate of any major U.S. city.[22]

It is possible, however, that Seattle's current success could be tomorrow's crisis. Given the long lead times to obtain permits and build new landfills or incinerators, the city has placed most (if not all) of its eggs in the recycling basket. Will Seattle officials prove to be farsighted prophets or merely wishful thinkers? Only time will tell.

Landfills. Public opposition to siting new landfills is the major impediment to integrating this disposal method into local MSW handling systems. One of the key points escaping public attention is that siting new landfills is virtually a must to avoid a solid waste crisis.

Even if source reduction, recycling, and waste-to-energy methods were utilized to their maximum, landfills would be needed to handle the residue. Even in land-poor Japan, with its high recycling rates and high use of incinerators, nearly a quarter of the waste stream is landfilled at one of 2,411 sites.[23]

Landfill technology is not complicated, but neither is it inexpensive. Preliminary site development (land acquisition, etc.) for a 100 acre, 20-year-operating-life landfill costs between $2.5-3.0 million. Final development of a single-liner system requires another $43 million ($64 million for a double-lined landfill). Environmental management (leachate and methane collection and management systems) and post-closure monitoring (equipment to measure pollutants in the air, surface water and groundwater) can bring up the expense by $20 million. In total, a single-lined landfill can cost nearly $66 million and a double-lined one, $87 million.[24]

Another important point is that landfills have never been safer. Archaeologist William Rathje offers an interesting historical perspective on landfilling:

> The safe and efficient disposal of garbage has never been deemed a high enough end in itself by the professionals [in the United States]. The goal has always been to get rid of garbage and do something else—create energy, make fertilizer, retrieve precious metals. In the case of sanitary landfills, their proponents hoped not only to dispose of mountains of garbage but also to reclaim thousands of acres of otherwise "waste" land. . . . The ideal places for landfills, they argued, were the very places that most scientists now believe to be the worst places to put garbage: along rivers or in wetlands. It is in unlined landfills in places like these that, not surprisingly, the problem of chemical "leachates" has been shown to be a matter of grave concern.[25]

Summary

The options for handling municipal solid waste are relatively few: source reduction, recycling, incineration, and landfilling. By and large, they are not overly complex.

No single method can handle all the trash generated by a community. Indeed, what is needed is an integrated approach—some from column A and some from column B—to accomplish MSW management in a cost-effective and safe manner.

International comparisons of waste disposal methods should also include consideration of the peculiar circumstances that shape the mix of waste-handling techniques for each nation. For example, would Japan recycle to the extent that it does, or rely so heavily on incinerators, if it had the abundance of geologically safe landfill sites that we have in the United States?

Public policy should enable local decision makers to make wise choices, fitted to their particular geographic and demographic circumstances. The next section provides a brief overview of current public policy directed at the MSW problem.

The MSW Public Policy Environment

To more effectively deal with the problem of safely disposing of municipal solid waste, local, state, and federal responsibilities should mesh better than they do now. It is instructive to first observe what the various levels of government are doing before suggesting ways that better integrate public policy at all levels.

Local-Government Proposals

For the most part, local officials have responded to the MSW "crisis" by emphasizing recycling—either drop-off centers or curbside recycling. For instance, in 1988 only three New Jersey towns recycled plastics; now, 300 cities in that state do so.[26]

Bans on yard wastes or separate higher fees for collecting and composting lawn clippings and leaves are also typical local-government responses. Likewise, restrictions on dumping auto batteries, tires, major appliances, and household hazardous wastes are common.

Some communities offer drop-off opportunities once or twice a year for disposal of hazardous household wastes to reduce landfilling of caustic and toxic household products. An unfortunate side effect of these special collection efforts is that they can actually cause a short-term increase in the amount of household hazardous waste thrown into the trash. The attention created by the announcement of the special drop-off event causes people to become concerned about the hazardous materials in their homes, so they throw them in the garbage.[27]

Cities also have become fond of banning plastic containers or mandating that manufacturers recycle them. When industry groups opposed a pending Minneapolis city ordinance requiring that plastic containers be recycled by mid-1991, they were clobbered by hostile public sentiment. Thus far, however, fewer than 25 cities have adopted plastic bans (usually on polystyrene containers used by restaurants).[28]

The movement to ban plastic containers may have suffered a setback recently when a State of New York appeals court overturned a Suffolk County, Long Island, ban on plastic food containers and grocery bags. Oddly enough, the law was done in by the lack of a complete environmental impact statement. The point was, that before passing environmental laws, elected officials should evaluate the environmental impact of their proposals, including likely substitution effects—in this case, higher rates of paper product disposal.[29]

If local officials were taking a more integrated approach to solving the impending MSW capacity crunch, we probably would see more waste-to-energy plants and siting of landfills. To some degree, the problem is an educational as well as political one. Public officials and citizens need to be convinced that these methods are safe.

One major change taking place is the privatization of solid waste disposal. New federal regulations for landfills and incinerators are driving up costs beyond what most cities can afford.

State Initiatives

Some of the state activities in the area of municipal solid waste disposal are similar to local initiatives, but writ large. Requirements for separate collection of recyclable materials, bans on yard wastes or requirements that landfill facilities have separate composting operations, and prohibitions against landfilling of car batteries, tires, and waste oil are becoming part of state law. Deposit laws on aluminum cans have been on the books in many states for some time. Their original purpose was to reduce littering but now deposit laws are being considered for beverage containers to promote recycling.

Florida has placed a tax of 10 cents a ton on nonrecycled newsprint; Nebraska will ban disposable diapers beginning in 1993; and a total of nine states have banned nonreturnable beverage bottles. Several states have enacted legislation requiring that garbage bags and disposable diapers be made of degradable materials.[30]

Maine passed a law that effectively bans aseptic boxes—the little plastic and paper drink boxes that do not require refrigeration. Although these boxes are more compact than plastic containers, safer than glass and clearly a hit with consumers, they will be banned because they are not easily recycled.[31]

In some instances, grassroots movements are even acting to force greater state level involvement. An initiative in Massachusetts would have required that, by 1996, all packaging sold in the state be reusable, recyclable or partly made up of recycled materials. The initiative is not going to be on the ballot this year due to a technicality, but the petition garnered 120,000 signatures.[32]

Much of this activity can best be described as *ad hoc*. The lack of planning and forethought going into these legislative responses does not bode well for taking the integrated approach to waste handling needed to avert a real crisis. With plastics as the prime whipping boy, state and local authorities are "straining at gnats" (plastics make up less than 7 percent of municipal solid wastes) while "swallowing camels," particularly paper and paperboard products (composing 41 percent of MSW). Moreover, they are relying heavily on recycling and ducking the tough issues of siting landfills and waste-to-energy plants.

One area of activity peculiar to the states is the prohibition of out-of-state dumping of consumer wastes. Whether these bans on garbage "imports" can withstand challenges based on the commerce clause of the constitution is debatable. Interestingly enough, federal legislative proposals have been offered that would grant a state the right to prohibit the landfilling of out-of-state trash. Thus, Congress seems quite willing to abdicate its constitutional authority and allow states to restrict the free flow of commerce in this instance.

Federal Policy Prescriptions

Congressional Piecemeal Proposals. Many congressional proposals for dealing with municipal solid waste problems are as *ad hoc* as state responses. Bills have been introduced to ban interstate transportation of solid wastes, to allow tax credits for recycled newsprint, to outlaw backhauling of garbage in trailers used to carry foodstuffs to the East Coast, and so on.

The White House also expressed support for a proposal to tax "virgin materials"—paper, glass, plastics and metals. A tax based on weight would be levied at the point of production, thus improving the competitive position of recycled materials.

The critical issue, of course, is how high to set the tax. Theoretically, the tax should be high enough to cover the "social costs," the costs of disposal, for each type of material that could be recycled. As Paul Portney, an economist at Resources for the Future, observed, "Once you've set the tax that reflects these costs, if it turns out that recycling is still prohibitively expensive, then so be it."[33] However, the proposal envisions a single tax rate ($20 a ton has been mentioned) for all materials, which would not reflect differing costs of disposal of individual materials.

More Comprehensive Legislation. A variety of comprehensive changes to the Resource Conservation and Recovery Act designed to deal with MSW disposal problems were proposed in the House and Senate in 1990. One such bill, H.R. 3735, authored by Congressman Thomas Luken (D-Ohio), provides a good example.

The most striking aspects of the Luken bill (and S. 1113, Senator Baucus's bill) were the provisions for a much expanded federal role in MSW disposal. H.R. 3735 would have required that state plans be approved by EPA and that each state be self-sufficient with regard to solid waste disposal capacity. EPA would have been obligated to provide model guidelines for landfills, incinerators, and monofills for incinerator ash. State plans also would have to have conformed to federal requirements for handling household hazardous waste, tires, yard waste, and large household appliances.

Rather than using the "stick" of a threatened federal-funds cutoff for noncomplying states (a technique that is quite popular in many federal environmental statutes), the Luken bill would have prohibited any state that could not certify adequate MSW disposal capacity from transporting its solid waste to another state.

The Luken bill would also have made EPA's goal of 25 percent recycling of solid waste a requirement within four years. A goal of 50 percent was set for eight years after passage of the bill.

H.R. 3735 also required the states to develop public education programs about solid waste, especially recycling, to be taught in the public schools. Congressional proposals clearly envision a prime role for EPA, and they do not leave to chance how the EPA is to do Congress' bidding.

Principles for Positive Public Policy

The piecemeal and punitive approach to the garbage dilemma that is typical of current public policy is not difficult to improve upon. Two basic principles could provide a dramatically improved public policy climate.

Principle #1: MSW Disposal Is Primarily a Local Problem

The MSW Task Force Recommendations. In this regard, the February 1989 final report to the EPA by the Municipal Solid Waste Task Force provides a valuable framework for apportioning municipal waste disposal responsibilities among federal, state, and local authorities. The Task Force identified six objectives for a national agenda to solve the municipal solid waste dilemma:

1. Increase the waste planning and management information (both technical and educational) available to states, local communities, waste handlers, citizens, and industry, and increase data collection for research and development.
2. Increase effective planning by waste handlers, local communities, and states.
3. Increase source reduction activities by the manufacturing industry, government, and citizens.
4. Increase recycling by government and by individual and corporate citizens.
5. Reduce risks from municipal solid waste combustion in order to protect human health and the environment.
6. Reduce risks from landfills in order to protect human health and the environment.[34]

Much of the emphasis of these objectives and of the Task Force report is on voluntary efforts, with the EPA playing a key information-provider role. The Task Force suggests that EPA develop technical and educational guidance, data collection and research and development programs, and act as a national clearinghouse for this information. Furthermore, it recommends that the agency set up a "peer matching" program that puts experts in waste management in touch with communities in need of assistance.

In addition, the Task Force envisions a cooperative effort by all levels of government to track volumes and types of wastes. EPA would take the lead in conducting R & D in technical areas related to combustion, landfilling, recycling, and source reduction.

The MSW Task Force sees a prime role for states in collecting and evaluating local MSW disposal plans and in setting statewide goals for waste handling. In this regard, dispute-resolution methods to prevent stalemates in siting waste management facilities is another key state responsibility.

The role described for EPA in source reduction is likewise primarily one of information gathering and dissemination. The MSW Task Force's final report states, "Although federal economic incentives and regulatory approaches [to source reduction] may be useful in the future, EPA is not recommending their adoption at this time." [35]

Similarly, EPA's role in recycling is seen as limited to investigating market incentives and disincentives for recycling a variety of municipal solid wastes. In addition, EPA would encourage federal departments and agencies to use recycled materials by continuing to develop procurement guidelines for these materials (guidelines for paper and tires have already been completed).

The MSW Task Force recommends key regulatory responsibilities for EPA to establish performance standards for combustors and landfills. The Resource Conservation and Recovery Act (RCRA) encourages the development of state solid waste management plans (Subtitle D, Subchapter IV). Current law authorizes EPA to promulgate regulations containing criteria to classify types of sanitary landfills, to facilitate the closing or upgrading of existing open dumps, and to revise landfill guidelines. Conspicuously absent in current RCRA law is a national permitting scheme or sanctions on states not meeting federal guidelines.

Applying the Task Force Principles. Municipal solid waste, by definition, is essentially a local problem. The primary role for EPA should be information gathering and dissemination, as suggested by the MSW task force. Guidelines for combustors and landfills do not need to carry threats with them. Local citizens will put pressure on elected officials to make them safe. If facilities are constructed and operated according to proper, state-of-the-art methods, citizens have little to fear from landfills and incinerators. There is no demonstrated negligence on the part of local officials that makes it necessary for permit systems to be approved by the EPA.

State governments have an interest in determining adequate MSW management capacity to support their populations. A positive role for state officials would be to act as agents for resolving local disputes in siting new waste management facilities. The LULU (Locally Unacceptable Land Use) problem requires a broader perspective than local officials can normally muster.

In this regard, Wisconsin's process of siting waste facilities is often mentioned as a model. The State Waste Facility Siting Board has the power to arbitrate between municipalities and developers as long as the agreements meet the requirements of the state. Of the 103 facilities subject to the law, 30 cases did not require Siting Board negotiation procedures, 26 reached an agreement, 41 were still in process (as of October 1989), 5 were withdrawn and 1 was going to arbitration. Even so, siting takes three to five years to complete and sometimes requires economic compensation to the host communities or payments to property owners.[36]

Principle #2: Maximize the Use of Market Incentives

The second principle to improve MSW public policy prescriptions is to make maximum use of market incentives to address garbage woes. The greatest "market failure" connected with solid waste disposal is the distortion of market incentives resulting from municipal systems that charge flat fees for garbage collection. Cities that collect trash from households as a community service or that charge a single fee regardless of the amount of trash put out by the household are encouraging wastefulness. Consumers are not being made aware of the true disposal costs resulting from their buying and discarding practices.

Seattle's successful recycling program is fundamentally a story of the power of marketplace incentives to change behavior. Although the volunteer efforts involved in Seattle's success have gained a great deal of attention, the clear motivating factor is that residents are faced with a "pay-as-you-throw" system of trash hauling. Homeowners pay $10.70 a month for a 19-gallon mini-can but $31.75 for three full-size, 32-gallon garbage cans. Curbside recycling of plastic beverage bottles, glass, cans, newspapers, and other waste paper is free. Four out of five Seattle households recycle and 90 percent put out one or less cans of garbage a week. Basically, it is the high incremental cost for additional garbage cans that has placed Seattle residents at the top of the recycling chart.

Newsprint furnishes a good example of the problems created by treating recycling as an ideology rather than just another waste management method. Once the province of the Boy Scouts, now nearly everyone is getting into the act of collecting newspapers. Increasing the supply of recycled newspapers lowers its price but has little short-run effect on demand. In the long run, consistently low prices may well alter the economics of newspaper recycling, creating major new uses. But in the short run, the swings in prices result in added uncertainties for recyclers—for-profit and not-for-profit organizations—that threaten their economic viability.

Were it not for the moral mystique of recycling, it would be apparent that newspapers may be a bane for landfills but are a boon for waste-to-energy facilities. Why should it be considered morally wrong to use a renewable resource like newspapers to create energy? By encouraging increased recycling of newspapers we may be encouraging greater use of *nonrenewable* energy resources to produce power and creating other potential environmental problems from the de-inking process involved in recycling, hardly the environmental high ground.

Economic incentives also provide a way out of the NIMBY syndrome. Citizens do not want recycling centers, incinerators, and landfills in their backyards because they reduce property values and adversely affect other nonpecuniary aspects of home ownership—a quiet neighborhood, less traffic congestion, and the like. The simple (but not easy) solution is to compensate property owners for their losses. An indirect form of compensation is to provide a public benefit—a firehouse, a park or civic center, or road improvements. Or restitution can be more direct, such as a lump-sum payment, or payment of property taxes for the homeowners. The point is that those enjoying the benefits of the waste disposal facility should offset at least some of the costs that are imposed on the areas surrounding the facility.

Even so, economic incentives will not always guarantee success in siting new landfills or combustors. When Waste Management (the nation's largest solid waste disposal firm) offered $25 million to residents of Lake Calumet, Illinois, to expand a landfill, they were turned down.[37]

Conclusion

Americans are concerned about the impending garbage crisis and rightly so. But we cannot solve our household and industrial solid waste problems unless we squarely address the real issues rather than the garbage mythology. The source of the solid waste crisis is not the lack of safe disposal sites but the lack of acceptance of new sites for landfills and incinerators. Modern facilities present little risk to public health and are not particularly complex, but neither are they welcome.

Better information is a key ingredient to resolving MSW management problems. Current media attention is already causing some positive changes in individual and business behavior. However, much of this information is inaccurate and the extent of the risks involved in solid waste disposal exaggerated; in short, some garbage exposés are "talking trash."

Waste management must be viewed as an integrated system. Source reduction, recycling, incinerating, and landfilling should be considered complementary methods of handling MSW. The proportion of the local waste stream managed by each of these methods should be determined based upon the costs and benefits of each.

Ad hoc command-and-control approaches to resolving the municipal solid waste dilemma are of limited value. Product bans often raise consumer costs or reduce convenience with little or no environmental improvement.

A more positive public-policy climate could be created by focusing on two basic principles:

- Solid waste disposal is first and foremost a local problem.
- Market incentives should be used to the maximum extent possible.

The Municipal Solid Waste Task Force's outline for the respective roles for federal, state, and local officials is worthy of greater attention. The chief function of the EPA should be as an information provider and disseminator, including providing guidelines for safe MSW disposal practices. The states are the logical choice as arbiters of local disputes, while local governments can best design and implement solid waste disposal systems fitted to their particular circumstances.

Economic "sticks" and "carrots" can do much to improve the status quo. Municipalities are continuing to distort the waste disposal market by providing cross subsidies from one part of the city budget to trash-hauling activities. Charges per can and curbside recycling should be used to send proper signals to households so that their purchasing choices are affected by disposal costs.

Recycling should be viewed from an economic perspective rather than as an article of faith. Some recycling efforts may require initial subsidies (from the public sector or the private sector) before becoming economically viable. Many of these efforts, however, are driven by an almost religious devotion to recycling as an end in itself. But recycling is just one of four MSW management options that must be evaluated on the basis of economic and environmental harm to produce an integrated system.

America may be facing a crisis in municipal solid waste management but it is not one caused by lack of solutions. Recycling, incinerator, and landfill technologies are up to the task if we stop demanding zero-risk disposal of household wastes. Reducing the environmental mythology and hysteria is a necessary first step down a long road to insuring adequate capacity of safe and economical waste management for present and future needs.

Notes

1. *Facing America's Trash: What Next for Municipal Solid Waste?* (Washington, D.C.: Office of Technology Assessment, 1989), p. 3.
2. *Landfill Capacity in the Year 2000* (Washington, D.C.: National Solid Wastes Management Association, 1989), p. 1.
3. *Facing America's Trash*, p. 272.
4. William Rathje, "Rubbish!" *The Atlantic Monthly*, December 1989, p. 103.
5. *Landfill Capacity*, p. 3.
6. EPA Municipal Solid Waste Task Force, *The Solid Waste Dilemma: An Agenda for Action* (Washington, D.C.: U.S. Environmental Protection Agency, 1989), p. 6.
7. Ibid., p. 12.
8. *Facing America's Trash*, p. 4.
9. Rathje, "Rubbish!" p. 101.
10. *Facing America's Trash*, p. 79. Figures for Australia, Bulgaria, etc., are from less reliable United Nations and Organization for Economic Cooperation and Development sources.
11. Rathje, "Rubbish!" p. 101.
12. *Facing America's Trash*, p. 84.
13. *Public Attitudes Toward Garbage Disposal* (Washington, D.C.: National Solid Wastes Management Association, May 3, 1990), pp. 3-4.
14. *Solid Waste Dilemma*, pp. 16-17.
15. Stephen Goode, "Artists Draw the Line at Cadmium," *Insight*, July 16, 1990, p. 60.
16. *Facing America's Trash, p. 28.*
17. *Public Attitudes*, p. 6.
18. *Solid Waste Dilemma*, p. 18.
19. *Facing America's Trash*, pp. 220-21.
20. Ibid., p. 36.

21. Randolph B. Smith, "Cleaner Incinerators Drawing Less Fire," *Wall Street Journal*, May 31, 1990, p. B1.

22. Randolph B. Smith, "Aided by Volunteers, Seattle Shows How Recycling Can Work," *Wall Street Journal*, July 19, 1990, p. A1.

23. "Buried Alive," *Newsweek*, November 27, 1989, p. 70.

24. *Landfill Capacity*, p. 4.

25. Rathje, "Rubbish!" p. 103.

26. David Stipp, "Towns Come Around to Plastic Recycling," *Wall Street Journal*, April 18, 1990, p. B1.

27. David Stipp, "Toxic-Waste Collector, At Residents' Disposal," *Wall Street Journal*, June 8, 1990, p. B1.

28. Randolph B. Smith, "Minneapolis, Providing a Model, Prods Industry to Bury Hatchet on Recycling," *Wall Street Journal*, February 2, 1990, p. B1.

29. Dennis Hevesi, "Ban on Plastics in Suffolk County is Overturned," *The New York Times*, March 4, 1990, p. 28.

30. Carol Matlack, "Recycling Bandwagon," *National Journal*, September 30, 1989, pp. 2400-2401.

31. Margaret E. Kriz, "It's Not Easy Buying Green," *National Journal*, July 14, 1990, p. 1745.

32. David Stipp, "Recycling Drive Faces Another Go-Around," *Wall Street Journal*, July 23, 1990, p. B1.

33. David Wessel and Rose Gutfeld, "White House Hopes to Spur Recycling with New Taxes on Certain Materials," *Wall Street Journal*, January 3, 1990, p. A3.

34. *Solid Waste Dilemma*, pp. 24-25.

35. Ibid., p. 42.

36. *Facing America's Trash*, p. 344.

37. "Buried Alive," p. 70.

2

Global Warming: Do We Know Enough to Act?

S. Fred Singer

Global catastrophes have always held a special fascination for the human mind. Since ancient times, philosophers and theologians contemplated worldwide cataclysms. In recent years, the global environment has taken center stage and has engaged scientists, politicians, environmental activists, and the media.

The new wave of global environmental concern has focused on global climate warming. It extends beyond apocalyptic visions and intense hype from environmental groups and the media—with passive cooperation from the involved bureaucracies and scientists—to a political drive to establish international controls over industrial processes and business operations. The scientific base for such drastic action is uncertain and contentious. However, this type of ill-conceived policy is sure to stifle economic growth and, hence, reduce human welfare.

This essay describes many of the scientific uncertainties surrounding global warming and urges that the necessary research be completed before precipitous and costly governmental actions are taken for the sake of dubious benefits.

Climate warming, as a possible consequence of greenhouse effects, has emerged as the major environmental issue of the 1990s. The easing of international tension with the Soviet Union could make the greenhouse effect, along with other global environmental concerns, a leading foreign policy issue. The wide acceptance of the Montreal Protocol—which limits and rolls back the manufacture of chlorofluorocarbons (CFCs) to avert future changes in stratospheric ozone—has encouraged environmental activists to call for similar controls on carbon dioxide (CO_2) from fossil-fuel burning. At conferences in Toronto (1988), The Hague (1989), and Geneva (1990), they have expressed disappointment with the White House for not supporting immediate action.

But should the United States assume "leadership" in a campaign that could cripple our economy? Would it not be more prudent to assure first, through scientific research, that the problem is both real and urgent? My conclusions can be summed up in a simple message: *The scientific base for an enhanced greenhouse warming, due to human activities, is too uncertain to justify drastic action at this time.*

There is little risk in delaying policy responses to this century-old problem, since there is every expectation that scientific understanding will be substantially improved within a few years. Instead of panicky, premature, and likely ineffective actions that would only slow down, but not stop, the further growth of CO_2, we may prefer to use the same resources—several trillion dollars, by some estimates—to increase our economic resilience so that we can then apply specific remedies, if and as necessary. That is not to say that prudent steps cannot be taken now. Indeed, many kinds of energy conservation and efficiency increases make economic sense even without the threat of greenhouse warming.

The Scientific Base

The scientific base for greenhouse warming includes some facts, lots of uncertainty, and just plain ignorance—requiring more observations, better theories, and more extensive calculations. Specifically, there is consensus about the increase in so-called greenhouse gases in the earth's atmosphere as a result of human activities. There is some uncertainty about the strength of "sources and sinks" for these gases, *i.e.*, their rate of generation and the rates of removal. There is major uncertainty and disagreement about whether this increase has caused a change in the climate during the last 100 years. There is also disagreement in the scientific community about *predicted* changes as a result of further increases in greenhouse gases; the computer models used to calculate future climate are not yet good enough.

As a consequence of this "shaky science," we cannot be sure whether the next century will bring a warming that is negligible or one that is significant. Finally, even if there were to be a global warming and associated climate changes, it is debatable whether the consequences will be good or bad. Likely, we would get some of each.

Greenhouse Gases

It has been common knowledge for about a century that the burning of fossil fuels—coal, oil, and gas—increases the normal atmospheric content of carbon dioxide (CO_2). Furthermore, conventional wisdom predicts an *enhancement* of the natural greenhouse effect and a warming of the global climate as a result. Advances in spectroscopy in the last century have produced evidence that CO_2 (and other molecules made up of more than two atoms) absorb infrared radiation. Consequently, a buildup in CO_2 would impede the escape of heat radiation from the earth's surface. In fact, it is the greenhouse effect from *naturally* occurring carbon dioxide and water vapor that has warmed the earth's surface for billions of years. Without the natural greenhouse effect, ours would be a frozen planet without life.

The policy issue now is whether the nearly 30 percent increase in CO_2, mainly since World War II, calls for immediate and drastic action. According to the prevailing theory and taking into account increases in the other trace gases that produce greenhouse effects, we have already gone halfway to an effective doubling of greenhouse gases—something that cannot be reversed in a century or more. Thus, the theory says we are locked into a temperature increase of 1.5° to 4.5° Celsius (2.7° to 8.1° Fahrenheit). The average global temperature should now be increasing at the rate of 0.3°C per decade![1]

Precise measurements of the increase in the atmospheric CO_2 date to the International Geophysical Year of 1957-58. More recently it has been discovered that other greenhouse gases—*i.e.*, gases that absorb strongly in the infrared portion of the spectrum—have also been increasing. This is due, at least in part, to human activities. These gases currently produce a greenhouse effect nearly equal to that of CO_2, but could soon outdistance carbon dioxide's greenhouse effect. A brief list of these non-CO_2 greenhouse gases follows:

- Methane is produced in large part by sources that relate to population growth. Among these sources are rice paddies, cattle, and oil-field operations.[2] Methane, now 20 percent of the greenhouse gas effect but growing twice as fast as CO_2, has more than doubled since pre-industrial times. It would become the most important greenhouse gas if CO_2 emissions were to stop.
- Nitrous oxide has increased by about 10 percent, most likely because of soil bacterial action promoted by the increased use of nitrogen fertilizers.
- Ozone from urban air pollution adds about 10 percent to the global greenhouse effect. It may decrease in the United States as a result of clean air legislation but increase in other parts of the world.
- Chlorofluorocarbons (CFC's or "freons"), manufactured for use in refrigeration, air conditioning, and industrial processes, could make a significant contribution but will soon be replaced by less-polluting substitutes.[3]
- The most effective greenhouse gas by far is water vapor.

The last item on the list, water vapor, is not a manmade greenhouse gas, but is assumed to amplify the warming effects of the manmade gases. It is not really known whether water vapor has increased in the atmosphere or whether it will increase in the future—although that is what all the model calculations assume. However, predictions of future warming should depend not only on the amount but also on the horizontal and especially the vertical distribution of water vapor, and on whether it will be in the atmosphere in the form of a gas, as liquid cloud droplets, or as ice particles. The current computer models are not refined enough to test these crucial points.

The Climate Record: The IPCC Report

Has there been a climate effect caused by the sharp increase in greenhouse gases during the last few decades? The data are ambiguous, to say the least. Advocates for immediate action profess to see a global warming of about 0.5°C since 1880. (See Figure 2.1.) They point to record temperatures experienced in

FIGURE 2.1
Changes in Northern Hemisphere Land and Marine Air Temperatures Since 1860

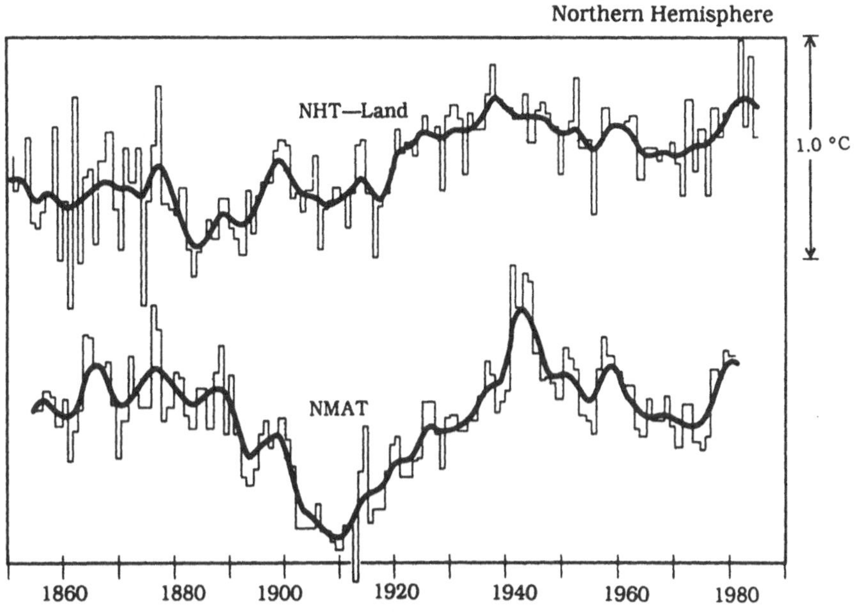

Note: This figure shows a major increase in temperature before 1938 (when greenhouse gas concentrations were low), a sustained *cooling* to 1975, and then a sharp rise to 1980 (more rapid than predicted by greenhouse models).

Source: P. D. Jones, T. M. L. Wigley, P. B. Wright, *Nature*, 322:430, 1986.

the 1980s and confidently predict a warming of as much as 5°C in the next century.

A United Nations-sponsored Intergovernmental Panel on Climate Change (IPCC) has been laying the groundwork for an international convention aimed at averting such a climate catastrophe. Its scientific base is a "Policymakers Summary" on greenhouse warming, released in June 1990,[4], said to represent a "scientific consensus." Far from it! The summary ignores valid scientific objections to the theoretical calculations that predict a global warming. It is silent about other human activities, notably the emission of sulfur dioxide in industrial processes, that are thought to promote a climate *cooling*. It plays fast and loose with the historical climate data that clearly disagree with the standard greenhouse warming theory. It also puts a "spin" on its major conclusions that can only serve to mislead the non-scientist decision makers who are earnestly seeking answers to global problems.

For example, to claim "certainty"—as the IPCC summary does—that "there is a natural greenhouse effect which already keeps the Earth warmer than it would otherwise be"[5] is on par with revealing that we know for sure that the earth is round. The greenhouse effect has been known for over a century and studied intensively for several decades. And to claim certainty that emissions of gases from human activities—carbon dioxide, methane, chlorofluorocarbons, nitrous oxide—"will enhance the greenhouse effect,"[6] is not particularly startling either, unless one can show that the additional warming is really significant.

Yet these two conclusions, issued without any further qualifications, must surely suggest to the unwary a future in which the United States heartland is turned into a parched desert by near-continuous droughts, its coasts lashed by frequent hurricanes or—worse still—flooded by oceans as sea levels rise to unprecedented heights. These images are assiduously promoted by many politicians and environmental groups and reported uncritically by the media, anxious to exaggerate cataclysmic disasters. They are not, however, based on scientific fact, and are not even supported by the details in the bulk of the IPCC report itself.

The IPCC report is based on faith in existing mathematical models that have not been able to "hindcast" the temperature changes experienced in the past century and, furthermore, have been in a state of flux. Within the last two years, in order to match more closely what is happening in nature, the predictions of the global-warming models have generally been scaled downward by including a more sophisticated accounting of cloud and ocean current effects.[7]

The report's claim to scientific consensus does nothing to enhance its credibility. Nevertheless, former British Prime Minister Margaret Thatcher, while still in office, latched on to it as a means of demonstrating her environmental credentials. She announced that she was joining ranks with other European leaders and had broken ranks with the Bush Administration by pledging that Britain would eliminate growth in carbon dioxide emissions over the next 15 years—provided other industrialized countries would do the same. Leaving herself this large loophole, she further stated that she was prepared to recommend a 30 percent reduction "in the currently projected levels of future carbon dioxide emissions." But since her projection included a healthy economic growth rate, the reduction amounts to nothing more than a leveling of emissions at about the present level.

White House officials have been more cautious, with Michael R. Deland, chairman of the president's Council on Environmental Quality, stating that the IPCC report was not the final word. White House Science Advisor D. Allan Bromley did not think that the report would have a major effect on the administration's actions. However, there are now several bills introduced in Congress which would have the United States restrict the use of energy—unilaterally, if necessary—to reduce the emission of carbon dioxide from fossil fuel burning.

The Other View of Climate Change

Many scientists do not accept the IPCC conclusions and call attention to the fact that the strongest temperature increase occurred *before* the major rise in greenhouse gas concentration. Figure 2.1 shows that this increase in both Northern Hemisphere land and marine air temperatures was followed by a 35-year-long temperature decrease, between 1940 and 1975, when concern arose about an approaching ice age! Following a sharp increase in average temperatures during 1975-1980, there has been no increase during the 1980s—in spite of record increases in greenhouse gases. Similarly, global *atmospheric* (rather than surface) temperatures, as measured by weather satellites, show no trend in the last decade.

Climatologists from the National Oceanographic and Atmospheric Administration (NOAA) find no overall warming in the U.S. temperature record—contrary to the findings from the rather poor-quality global record.[8] Using a technique that eliminates urban "heat islands" and other local distorting effects, they confirm the temperature rise before 1940, but then show a general decline. (See Figure 2.2.) Reginald Newell and colleagues at MIT report no substantial change in the global sea surface temperature in the past century.[9] This is important because the ocean, with its much greater heat inertia, should control any atmospheric climate change.

Perhaps most significant are the studies that document a relative rise in nocturnal temperatures in the United States in the last 60 years, while daytime values stayed the same or de-

FIGURE 2.2
U.S. Temperature Record Since 1900

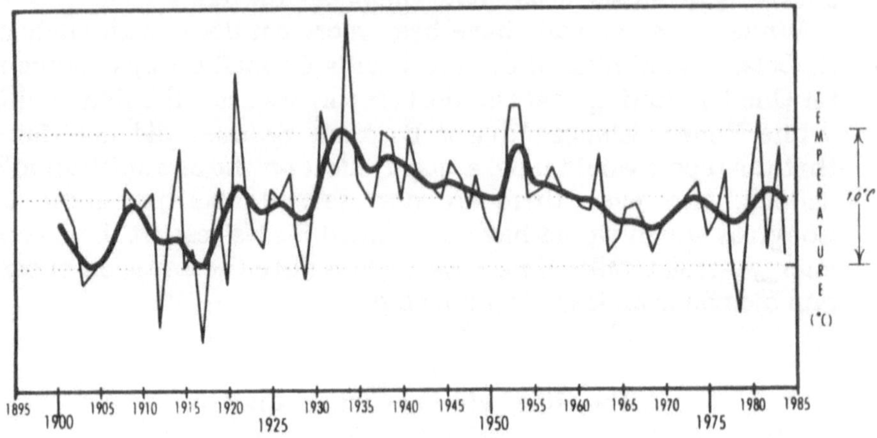

Note: This figure shows the annual average and smoothed temperatures, weighted and corrected for "urbanization" effects.

Source: Data from T. R. Karl, et al., *Historical Climatology Series 4-5* (Asheville, N.C.: NOAA National Climate Data Center, 1988).

clined.[10] This is just what one would expect from the increase in atmospheric greenhouse gas concentration. But its consequences, as Patrick Michaels of the University of Virginia has pointed out, are benign: a longer growing season, fewer frosts, and no increase in soil evaporation.[11]

The most precise measurement, free of local influences on temperature gauges, and truly covering the whole globe, including the oceans, comes from a microwave experiment in satellites.[12] The microwave data measure the temperature of the lower atmosphere itself—globally, continuously, both day and night, and with high precision. They show essentially no net temperature change during the past ten years. The time span, however, is still too short to draw meaningful conclusions. But data over the next decade should allow us to test the IPCC projection that global temperatures will rise by about 0.3°C per decade on average, resulting in a global increase of 2°C by 2025 and 6°C by the end of the twenty-first century.[13]

It is, therefore, fair to say that we have not seen the huge greenhouse warming of about 1.5°C expected by now from some theories and played up by the media. *But why not?* This scientific puzzle has many suggested solutions:

- The warming has been "soaked up" by the ocean and will appear after a delay of some decades.[14] This is plausible, but there is no evidence to support this theory until deep-ocean temperatures are measured on a routine basis.
- The warming exists as predicted, but has been hidden by offsetting climate changes caused by volcanoes, solar variations, or other natural causes as yet unspecified—such as the cooling from an approaching ice age.[15] (See box and Figure 2.3.)
- The warming has been overestimated by the existing models. Meteorologists Hugh Ellsaesser of the Livermore National Laboratory and Richard Lindzen of MIT believe that the models do not take proper account of tropical convection and thereby overestimate the amplifying effects of water vapor.[16] Other atmospheric scientists suggest that the extent of cloudiness may increase as ocean temperatures rise and as evaporation increases. Since clouds reflect incoming solar radiation, the resultant cooling could offset much of the greenhouse warming.[17]
- Most intriguing has been the suggestion by British researchers that sulfates from smokestacks—the precursors of acid rain—may have played a role in producing an increase in bright stratocumulus clouds, thereby offsetting greenhouse warming. Climate records are consistent with this hypothesis, leading to the surprising conclusion that cleaning up smokestack emissions could enhance warming, at least in the northern hemisphere.[18]

Each hypothesis has vocal proponents—and opponents—in the scientific community. The jury is out until better data become available.

Mathematical Models

Indeed, there is much to complain about when it comes to predictions of future climate; but there is really no alternative to global climate models. Half a dozen of these General Circulation Models (GCM) are now running, mostly in the United States.

Historical Climate Changes

Global temperatures have been declining since the dinosaurs roamed the earth, some 70 million years ago. About two million years ago a new "ice age" began—most probably as a result of the drift of the continents and the build-up of mountains. Since that time the earth has seen at least 17 cycles of glaciation, interrupted by short (10,000 to 12,000 years) interglacial (warm) periods. We are now in such an interglacial interval, the Holocene, that started 10,800 years ago; the onset of the next glacial cycle cannot be very far away. (See Figure 2.3.)

It is believed that the length of a cycle, about 100,000 to 120,000 years, is controlled by small changes in the seasonal and latitudinal distribution of solar energy received as a result of changes in the earth's orbit and spin axis—according to a theory first suggested by the Yugoslav astronomer Milankovitch. While the theory can explain the timing, the detailed mechanism is not well understood—especially the sudden transition from full glacial to interglacial warming. Very likely, an ocean-atmosphere interaction is triggered and becomes the direct cause of the transition in climate.

The climate record also reveals evidence for major climatic changes on time scales shorter than those for astronomical cycles. During the past millennium the earth experienced a "climate optimum" around 1100 AD, when Vikings found Greenland to be green and Vinland (Labrador?) able to support grape growing. The "Little Ice Age" found European glaciers advancing well before 1600, and suddenly retreating starting in 1860. The warming of 0.5°C, reported in the global temperature record since 1880, may be the escape from this Little Ice Age rather than our entrance into the human greenhouse.

Even though these computer programs use similar basic atmospheric physics, they give different results.

There is general agreement among modelers that there should be some global warming, but the actual estimates vary widely. Models predict that if the effective greenhouse gases double, the

FIGURE 2.3
Global Climate Change over
Different Time Scales

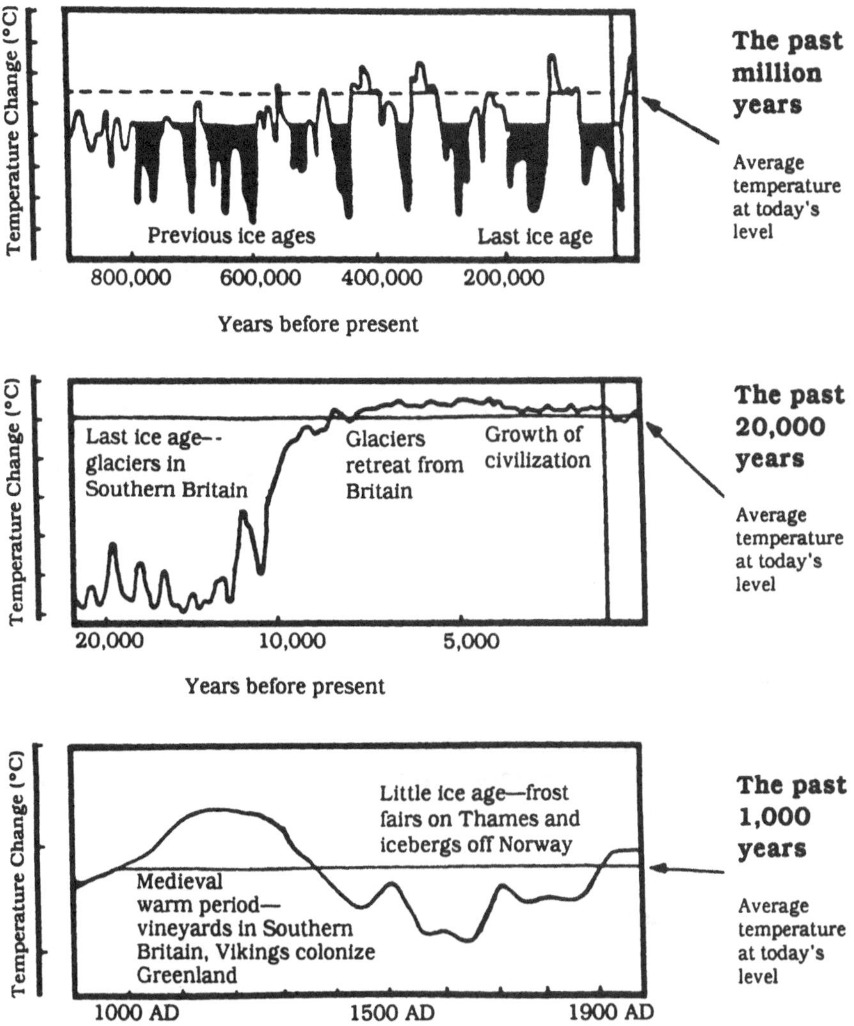

Source: J. T. Houghton, G. J. Jenkins, and J. J. Ephraums, eds., *Climate Change: The IPCC Scientific Assessment* (Cambridge: Cambridge University Press, 1990), p. 202.

average global temperature should increase between 1.5° and 4.5°C. (These projections of global warming were unchanged for many years, then crept up, and have recently dropped back. During 1989, some modelers cut their predictions in half as they tried to better model clouds and ocean currents.) There is also disagreement on the regional distribution of this warming and on where the increased precipitation will go. One British computer model forecast equal levels of rainfall for the Sahara Desert and Scotland.[19]

The models are "tuned" to give the right mean temperature and seasonal temperature variation, but they fall short of modeling other important atmospheric processes, such as the poleward transport of energy. Nor do they routinely encompass longer-scale processes that involve the oceans or the ice and snow in the earth's cryosphere. Fine-scale processes that involve convection, cloud formation, boundary layers, or the earth's detailed topography are inadequately incorporated into the models.

There are also serious disagreements between model results and the actual experience from the detailed climate record of the past century. The "fingerprints" characteristic of greenhouse warming are not to be found in the record. Some of the expected results of greenhouse warming and the actual findings are listed below:

- Existing models predict a strong warming of the polar regions and of the tropical upper atmosphere—all contrary to observations.
- Contrary to model predictions, observations suggest that warming has been more rapid over the oceans than over the land, and over the southern than over the northern hemisphere.
- As noted above, the bulk of the warming occurred before 1940, before more than a quarter of the fossil fuel CO_2 had been released into the atmosphere and before half of the biospheric release of CO_2 had occurred. Yet during the period of 1940 to 1975, cooling occurred in spite of the rapid release of CO_2 from fossil fuel burning.
- The bulk of the warming occurred in two short bursts, around 1920 and from 1975 to 1980. The models give us no ready explanation of these occurrences.[20]

According to climatologist Stephen Schneider: Models are better than "handwaving."[21] But not much better! There is hope, however, that research, including satellite observations and ocean data, will provide many of the answers to these greenhouse puzzles within this decade. And faster computers will have higher resolution and incorporate the detailed and more complicated interactions that are now neglected.

Impacts of Climate Change

Based on extrapolation of the climate record, the most likely outcome of the buildup in greenhouse gases is a modest warming of less than 1°C (average) in the next century, with the increase concentrated at high latitudes and in the winter. Is this necessarily a bad result? One should perhaps recall that only a decade ago, when climate cooling was a looming issue, a U.S. government study calculated a huge national cost associated with such cooling.[22] More to the point perhaps, actual climate cooling, experienced during the Little Ice Age (see box on page 38) or in the famous 1816 New England "year without a summer," caused large agricultural losses and even famines.

It would seem that if cooling is bad, then warming should be good. This statement is predicated on warming being slow enough so that adjustment is easy and relatively cost-free. Even though crop varieties are available that can benefit from higher temperatures with either more or less moisture, the soils themselves may not be able to adjust as quickly.[23] Agriculturalists expect that with increased atmospheric CO_2—which is, after all, plant food—plants will grow faster and need less water. The warmer night temperatures suggested by NOAA climatologist Tom Karl's research translate to longer growing seasons and fewer frosts. Increased global precipitation should also be beneficial to plant growth.[24]

It is important to keep in mind also that year-to-year changes at any location are far greater and more rapid than what might be expected from greenhouse warming. Nature, crops, and people have already adapted to such large short-term swings. It is the *extreme* climate events—crippling winters, persistent droughts, killer hurricanes, and the like—that cause the great ecological and economic problems. But there is no indication from modeling or from actual experience that such extreme events would become more frequent if greenhouse warming ever becomes appreciable. The exception might be tropical cyclones, which, according to Robert Balling and Randall Cerveney of Arizona State University, would be more frequent but weaker, cooling vast areas of the ocean surface and increasing annual rainfall by 10 to 15 percent.[25] Climate models predict that polar temperatures should warm the most, thus reducing the driving force for severe winter weather events.

Finally, there is the fear of sea-level rise and catastrophic flooding as glaciers melt. The cryosphere contains enough ice to raise sea level by 100 meters. Conversely, during recent ice ages enough ice accumulated to drop sea level 100 meters below the present level. These are extreme possibilities. For example, tidal gauge records of the past century suggest that sea level has risen modestly, by about 0.1 to 0.2 meters.[26] But the gauges measure only *relative* sea level, and many of their locations have dropped because of land subsidence. Besides, the locations are too highly concentrated geographically (mostly on the U.S. east coast) to permit global conclusions. The situation will improve greatly, however, in the next few years as precise *absolute* global data become available from a variety of satellite systems.

In the meantime, satellite radar-altimeters have already given a surprising result. As reported by NASA scientists, Greenland's ice sheets are gaining in thickness.[27] A net increase in the ice stored in the cryosphere and an inferred *drop* in sea level leads to somewhat different predictions about future sea level. It is clearly important to verify these results by other techniques and also to get more direct data on current sea-level changes.

Summarizing the available evidence, we conclude that even if significant warming were to occur in the next century, the net impact may well be beneficial. This conclusion would be even more compelling if the long-anticipated ice age were on its way.

What to Do?

In view of the uncertainties about the degree of warming, and the even greater uncertainty about its possible impact, what should we do? During the time that an expanded research program reduces or eliminates these uncertainties, we can implement policies and pursue approaches that make sense *even if the enhanced greenhouse effect does not exist.* A variety of sensible actions suggest themselves.

Energy Conservation

Global energy conservation can best be achieved by pricing rather than by command-and-control methods. Prices should include the external costs that are avoided by the user and transferred to someone else. The idea is to have the polluter or the beneficiary pay the cost. An example is peak pricing for electric power. Yet another example, appropriate to the greenhouse discussion, is to increase the tax on gasoline to make it a true highway user fee—instead of the current method of having most capital and maintenance costs paid by various state taxes. Congress has lacked the courage for such a direct approach, preferring instead regulation (such as the Corporate Average Fuel Economy standards) that is mostly ineffective and produces large indirect costs for the consumer.

Improve Energy Efficiency

Energy efficiency should be attainable without much intervention, provided it pays for itself. A good rule of thumb is: If it isn't economical, then it probably wastes energy in the process and we shouldn't be doing it. Over-conservation can waste as much energy as under-conservation. But provided that energy is properly priced, the job for government is to remove the institutional and other road blocks. Helpful government actions include, but are not limited to, the following:

- Providing information to consumers, especially on life-cycle costs for home heating, lighting, refrigerators, and other appliances.
- Encouraging turnover and replacement of older, less efficient (and often more polluting) capital equipment, such as cars, machinery, and power plants. Some existing policies that make new equipment too costly run counter to this goal.
- Stimulating private-sector development of more efficient systems, such as combined-cycle power plants or a more efficient internal combustion engine.

Encourage Use of Non-Fossil Fuels

Nuclear power is competitive now, and in many countries is cheaper than fossil-fuel power, yet is often opposed on environmental grounds. The problems cited against nuclear energy, such as disposal of spent nuclear fuel, are mostly political and ideological rather than technical. Nuclear energy from fusion rather than uranium fission may be a longer-term possibility, but the time horizon for this development is uncertain.

Solar energy and other forms of renewable energy should also become more competitive as their costs drop and as fossil-fuel prices rise. However, solar energy is both highly variable and very dilute. In fact, it takes a football field of solar cells in a sunny climate to supply the total energy requirements of the average U.S. household. Wind energy and biomass are other forms of energy that are competitive in certain applications. Schemes to extract energy from temperature differences in the ocean have been suggested as inexhaustible sources of non-polluting hydrogen fuel, provided we can solve the daunting technical problems.

Other Futuristic Solutions

If greenhouse warming ever becomes a real problem, there are numerous proposals for removing CO_2 from the atmosphere—that is, increasing CO_2 "sinks." Reforestation is widely talked about, but probably not cost-effective. However, natural expansion of boreal forests in a warming climate would sequester atmospheric CO_2. A novel idea, proposed by California oceanographer John Martin, is to fertilize the Antarctic Ocean and let plankton growth do the job of converting CO_2 into biomaterial. The limiting trace nutrient may be iron, which could be supplied and dispersed economically.[28]

And, if all else fails, there is the possibility of putting "venetian blinds" satellites into earth orbit to modulate the amount of sunshine reaching the earth. These satellites could also generate electric power and beam it to Earth, as originally suggested by Peter Glaser of the A. D. Little organization.[29] Such a scheme may sound farfetched—and possibly it is—but many other futuristic projects, like covering the Sahara with solar cells or Australia with trees, have been discussed seriously.

What Not to Do About Greenhouse Warming

Environmental groups and congressmen have renewed demands for immediate action on greenhouse warming. A key point of the several bills already introduced is for the United

States to cap and reduce—unilaterally, if necessary—the emission of the major greenhouse gas, carbon dioxide resulting from fossil-fuel burning. The White House would be well advised to resist such pressure to place scientifically arbitrary and economically ruinous limits on energy generation.

The Hidden-Agenda Problem

It should come as no surprise to anyone that many people are hyping the greenhouse warming "threat" to push their own pet agendas. They seem undeterred by the growing scientific evidence that shows no climate effects from the increase in atmospheric greenhouse gases. For example, global temperatures did not increase during the past decade—contrary to cataclysmic predictions. Yet the *Today* show, PBS-TV specials like "Crisis in the Atmosphere," and most of the print media all still preach impending doom in the form of the collapse of global agriculture or a catastrophic rise in sea levels. An editor of *Time* magazine even assures journalists that it is all right to become environmental advocates, never mind scientific facts.[30]

Why do so many different groups focus on greenhouse warming? Because this issue provides a wonderful excuse for doing things that they already want to do, under the guise of saving the planet. We find in one corner proponents of nuclear energy (which emits no CO_2), who see a chance to refurbish their public image. Next to them are natural gas producers, keen on beating out competition from cheaper, but more polluting coal. Even scientists are becoming cheerleaders: budgets for climate research just jumped to over one billion dollars.

In another corner we find proponents of energy conservation and renewable energy. These are quite commendable goals, really, except for those uneconomical measures that waste more energy than they save. The extremists in this crowd oppose *all* energy growth and economic growth.

More dangerous are those who have a hidden *political* agenda, most often oriented against business, the free market, and the capitalistic system. Of course, after the collapse of socialism in Eastern Europe it is no longer fashionable to argue for state ownership of industrial concerns. The alternative is to *control* private firms by regulating every step of every manufacturing process.

And then there are those who visualize global warming as a vehicle for international action, preferably with lots of treaties and protocols to control CO_2 or perhaps even methane. Some view the greenhouse effect as a launch platform for an ambitious foreign aid program, others as the justification for global multi-billion dollar afforestation projects, while still others would use it to encourage (or even enforce) global population control.

Ineffective and Costly Policies: The Carbon Tax

With all this private-sector support, there is little reason why the U.S. Congress should exempt itself from all the popular excitement that bids to take over much of domestic and even foreign policy. Let's look at policies that should *not* be established—most of them enshrined in current legislative proposals.

The notion of capping CO_2 emissions, or even rolling back emissions by 20 to 50 percent, is based on the facile idea that energy conservation can be achieved by decree. It is often combined with the idea of a "carbon tax" to make it sound like a "free-market" proposal. If enacted unilaterally—in order to demonstrate what some refer to as "U.S. leadership" and others as economic suicide—it would hamper U.S. economic growth. The impact would fall mainly on the poorer segment of the population, like so many pollution control proposals.

If applied to the rest of the world, capping CO_2 emissions would certainly be denounced as "eco-imperialism," a scheme to stop economic development in the less-developed countries (LDCs). The billions of people in the LDCs want cars, refrigerators, TVs, and air conditioning. It might take a large bribe to make them go along with any scheme that would deny them the quality of life that comes with the use of energy and electricity.

A carbon tax, imposed primarily on coal, has many strikes against it:

- Coal is used mainly to produce electric power. Since public utilities would pass the increased fuel cost along to ratepayers, a carbon tax amounts to little more than a regressive value-added tax.
- A carbon tax is not only unfair but also counterproductive to the national goal of reducing oil imports. Cutting the use of domestically produced coal would also give a windfall to oil and gas producers (read: OPEC), at the expense of consumers.

Of course, all this exclusive emphasis on limiting energy generation and coal use is misplaced and irrational, for a number of reasons:

- CO_2 is added to the atmosphere by activities other than energy generation. Deforestation and soil erosion, primarily in the LDCs, are believed to provide 25 percent or more of the CO_2 in the atmosphere. By contrast, U.S. forest inventories have been growing since 1920, sequestering carbon dioxide from the atmosphere.[31] A further uncertainty is the fraction of CO_2 that is absorbed into the ocean. Estimates run as high as 50 percent. Until these numbers are made more precise, it is difficult to predict the result of any remedial policy.
- The United States is only a modest contributor to atmospheric carbon dioxide from fossil-fuel burning, currently less than 25 percent. More importantly, this percentage is shrinking rapidly as LDCs increase both their populations and their standards of living. According to Worldwatch Institute, a widely cited environmental think tank, CO_2 contributions from LDCs have grown rapidly in the last decades and can be expected to explode in the coming century. Between 1960 and 1987, carbon dioxide emissions from fossil fuels increased by 54 percent for the United States, 26 percent for France, 22 percent for West Germany, 161 percent for the USSR, 292 percent for Japan, 176 percent for China, 307 percent for Brazil, and 357 percent for India—while the United Kingdom's emissions actually decreased by 3 percent.[32]
- Most important, carbon dioxide will be losing its pre-eminent position as a greenhouse gas to other gases, like methane, for which no control strategy has been developed. How does one control emissions from rice paddies and cows, which are among the primary sources of methane? Currently, CO_2 contributes over 50 percent of the greenhouse effect, but carbon dioxide is becoming "saturated," *i.e.* an increase in its atmospheric concentration no longer produces a proportional increase in the effect.

The upshot of all these considerations is that even our most drastic measures to limit fossil-fuel burning can do little to stave off an inevitable rise in atmospheric greenhouse gases. Doubling of these gases could be delayed from perhaps 2040 to only 2045 but would cost our economy a trillion dollars by some estimates.[33]

Reflections and Conclusions

As stated before, the available evidence indicates that the net impact of significant warming in the next century may well be beneficial. Yale economist William Nordhaus, one of the few who has been trying to deal quantitatively with the economics of the greenhouse effect, has pointed out that ". . . those who argue for strong measures to slow greenhouse warming have reached their conclusion without any discernible analysis of the costs and benefits."[34]

Those who are charged with policy decisions that can affect economic growth and the welfare of billions of people would do well to move cautiously and insist on a thorough understanding of the physics of the atmospheric greenhouse before taking hasty and far-reaching actions that would drive energy prices sky-high. Stringent controls enacted now would be economically devastating without being able to affect greatly the growth of greenhouse gases in the atmosphere.

Energy conservation, efficiency increases, and use of non-fossil fuels are all prudent policies, as long as they are cost-effective. But more drastic, precipitous—and especially, unilateral—steps to delay the putative greenhouse impacts can cost jobs and economic prosperity without being effective. The reality of the greenhouse effect has been around and known for over a century. It would be wise to complete the ongoing and recently expanded research so that we will know what we are doing before we act. "Look before you leap" is still good advice.

Notes

1. J. T. Houghton, G. J. Jenkins, and J. J. Ephraums, eds., *Climate Change: The IPCC Scientific Assessment* (Cambridge: Cambridge University Press, 1990).
2. R. J. Cicerone, "Methane in the Atmosphere," in S. F. Singer, ed., *Global Climate Change: Natural and Human Influences* (New York: Paragon House Publishers, 1989).
3. F. S. Rowland, "Chlorofluorocarbons, Stratospheric Ozone, and the Antarctic Ozone Hole," in Singer, ed., *Global Climate Change.*

4. Houghton, Jenkins, Ephraums, eds., *IPCC Scientific Assessment*, p. xi.

5. Ibid.

6. Ibid.

7. P. J. Michaels, "Greenhouse Disaster?" submitted to *Science* (1990).

8. T. R. Karl et al., *Historical Climatology Series 4-5* (Asheville, N.C.: NOAA National Climate Data Center, 1988); also T. R. Karl et al., in *Journal of Climate and Applied Meteorology* 23:1489, 1984.

9. R. E. Newell et al., in *Geophysical Research Letters* 16:32, 1989; M. Bottomly et al., *Global Ocean Surface Temperature Atlas* (Cambridge: MIT Department of Earth and Planetary Science, 1990).

10. Karl et al., *Historical Climatology*.

11. Michaels, "Greenhouse Disaster?"

12. R. W. Spencer and J. R. Christy, *Science* 247:1558, 1990.

13. Houghton, Jenkins, Ephraums, eds., *IPCC Scientific Assessment*.

14. K. Bryan et al., *Science* 215:56, 1982; W. M. Washington and G. A. Meehl, *Climate Dynamics* 4:1, 1989.

15. S. F. Singer, "Postscript" in *Global Climate Change*.

16. H. W. Ellsaesser, *Atmospheric Environment* 18:431, 1984; R. J. Lindzen, *Bulletin of the American Meteorological Society* 71:288, 1990.

17. S. F. Singer, "Fact and Fancy on Greenhouse Earth," *Wall Street Journal*, August 30, 1988.

18. Michaels, "Greenhouse Disaster?"

19. In *The Greenhouse Conspiracy*, produced by TVF Productions, London. Cited in "Warming Up to the Facts," *Wall Street Journal*, January 11, 1991, p. A10.

20. H. W. Ellsaesser, in *Symposium on Global Climate Change*, July 30-31, 1990, Los Angeles, published by the California Energy Commission.

21. S. M. Schneider, "Climate Modeling," *Scientific American* 256:72-80, May 1987.

22. R. d'Arge et al., in *CIAP Monograph 6*, DOT-TST-75-56 (Washington, D.C.: U.S. Department of Transportation, 1975).

23. S. F. Singer, "The Responses to Climate Change: The Significance of Different Time Scales," in *Living with Climate Change* (McLean, Va.: MITRE Corp., 1977).

24. S. B. Idso, *Carbon Dioxide and Global Change: Earth in Transition* (Tempe, Ariz.: IBR Press, 1989); W. E. Reifsnyder, in *Agricultural and Forest Meteorology* 47:349, 1989.

25. R. C. Balling and R. S. Cerveney, in *Journal of Applied Meteorology* 27:881, 1988.

26. Houghton, Jenkins, Ephraums, eds., *IPCC Scientific Assessment*, p. 261.

27. H. J. Zwally, in *Science* 246:1589, 1989.

28. J. Martin et al., in *Nature* 345:516, 1990.

29. P. E. Glaser, "Power from the Sun: Its Future," *Science* 162:857, 1968.

30. D. Brooks, "Journalists and Others for Saving the Planet," *Wall Street Journal*, October 5, 1989.

31. C. Flavin, "Slowing Global Warming," *State of the World* (Washington, D.C.: Worldwatch Institute, 1990).

32. Ibid.

33. W. D. Nordhaus, in H. Aaron, ed., *Setting National Priorities: Policy for the Nineties* (Washington, D.C.: Brookings Institution, 1990), pp. 185-211; A. S. Manne and R. G. Richels, *Global CO_2 Emissions Reductions: The Impacts of Rising Energy Costs* (Palo Alto: Stanford University and Electric Power Research Institute, February 1990).

34. Nordhaus in *Setting National Priorities*.

3

Battling Smog

Anne Sholtz-Vogt and Kenneth Chilton

Amending the Clean Air Act was one of the top priorities of the 101st Congress and a subject of much debate from the fall of 1989 through the fall of 1990. A key portion of the comprehensive legislation focused on atmospheric ozone—urban smog. The cost estimates for these smog provisions range from $3 billion to $10 billion annually.

For the most part, the public debate on smog remains uninformed, despite all the recent attention. Few citizens know the true health risks associated with ozone, their exposure to "unhealthful" levels of the pollutant, or the costs and benefits of proposals to reduce smog.

What is still needed in the efforts to battle smog is a broader perspective; one that answers basic questions, such as:

- What does it mean to be a nonattainment area?
- What are the health effects of ozone pollution?
- What causes ozone buildup?
- What are the costs and benefits of reducing smog levels?
- Will the 1990 Clean Air Act amendments promote cost-effective progress toward a healthier environment?

What Does It Mean to Be a Nonattainment Area?

"Nonattainment" has a very precise meaning as defined in the Clean Air Act (CAA). If the fourth highest daily monitor reading taken during the most recent three-year period registers an ozone concentration above 0.120 parts per million (ppm) for more than one hour, an area is not in attainment. (The fourth highest daily maximum reading is referred to as the "design value.") The definition is cut-and-dried, taking no account of any other monitor readings in an area, which may be far less than the highest monitor reading.

According to Environmental Protection Agency (EPA) figures for 1987, 88.6 million Americans lived in counties that exceeded the CAA ozone standard for that year.[1] Table 3.1 is a list of "nonattainment" areas whose fourth-highest ozone monitor reading for the period 1986-1988 exceeded the 0.120 ppm standard by 33 percent or more. (The table is in the order of the severity of the ozone problem—in "design value" order.)

It is possible that the EPA may decide to classify areas using 1987-1989 as the relevant base years. Nonetheless, the results in Table 3.1 would be largely unaffected because the unusually severe ozone conditions of 1988 would continue to have a determining effect on the design value.

The final column of Table 3.1 indicates the percent that emissions of volatile organic compounds (VOCs)—reactive hydrocarbons from gasoline, solvents, paints, and decaying vegetation—which are prime ingredients of ozone, must be reduced for a particular area to reach the 0.120 parts per million standard. One point is abundantly clear: many cities need to make large reductions in emissions to reach attainment. More than 63 million Americans currently live in areas where VOCs will need to be reduced by 40 percent or more (from 1985 emission levels) to meet the standard.

The simple designation of an area as nonattainment is misleading, however. This definition focuses on a one-hour peak concentration figure in order to provide an adequate margin of safety against *any* adverse health effects. But this is not necessarily the best measure of overall air quality.

For example, Los Angeles undoubtedly suffers from the worst smog problem in the United States. L.A.'s design value is 0.34 ppm, nearly three times the air quality standard. Area ozone levels at the highest monitor exceeded 0.120 ppm for one hour or more for an average of 145 days a year during the period 1986-1988.[2]

But a different picture of the Los Angeles ozone problem forms, based on an alternative set of facts. When the average of all monitors over all hours is considered, readings were above

TABLE 3.1
Urban Emissions Reductions Needed to Reach Ozone Attainment

Area	Population (millions)	1986 -1988 design value	Percent above standard	NMOC/ NO_x ratio	Required VOC reductions (%)[a]
Los Angeles, CA	8.3	0.34	183%	10.4	80%
New York City, NY	8.5	0.22	83	11.7	67
Chicago, IL	6.2	0.20	67	7.8	53
Houston, TX	3.2	0.19	58	10.9	60
Baltimore, MD	2.9	0.18	50	7.1	42
Hartford, CT	0.7	0.18	50	9.4	53
Milwaukee, WI	1.4	0.18	50	9.6	53
Muskegon, MI	0.2	0.18	50	12.2	61
Philadelphia, PA	4.5	0.18	50	7.2	43
Portsmouth-Dover, NH-ME[b]	0.2	0.18	50	12.2	38
San Diego, CA	2.4	0.18	50	7.4	44
Atlanta, GA	2.6	0.17	42	6.5	36
Boston, MA[b]	2.8	0.17	42	5.7	34
El Paso, TX	0.6	0.17	42	8.5	46
Fresno, CA	0.6	0.17	42	12.2	55
Huntington-Ashland, WV-KY	0.3	0.17	42	12.2	58
Louisville, KY	1.0	0.17	42	12.2	58
Parkersburg-Marietta, WV-OH	0.2	0.17	42	12.2	58
Sheboygan, WI	0.1	0.17	42	12.2	58
Worcester, MA	0.4	0.17	42	12.2	58
Bakersfield, CA	0.5	0.16	33	12.2	49
Baton Rouge, LA	0.5	0.16	33	14.8	60
Beaumont and Port Arthur, TX	0.4	0.16	33	22.4	72
Cincinnati, OH	1.4	0.16	33	9.1	44

TABLE 3.1 (continued)

Area	Population (millions)	1986 -1988 design value	Percent above standard	NMOC/ NO_x ratio	Required VOC reductions (%)[a]
Dallas-Ft. Worth, TX	2.4	0.16	33%	9.1	44%
Portland, ME	0.2	0.16	33	11.6	52
Providence, RI	0.6	0.16	33	12.2	54
Sacramento, CA	1.3	0.16	33	9.6	45
Springfield, MA	0.2	0.16	33	12.2	53
St. Louis, MO	2.4	0.16	33	10.7	50
Washington, D.C.	3.6	0.16	33	9.2	45

[a] This column shows an estimate of the percent that VOCs must be reduced for an area to reach the 0.12 ppm standard.

[b] Transport emissions are included in calculations of required reductions.

Source: Population figures are from EPA, *National Air Quality and Emissions Trends Report, 1987*, Table 4-3. Design values are from the July 27, 1989 EPA *Environmental News*. NMOC/NO_x ratios and information needed to calculate percentage reductions taken from a personal correspondence from Francis Bunyard, EPA Office of Air Quality Planning and Standards, August 1989.

the standard less than three percent of the time in the five-year period 1981-1985. Further, average readings at all monitors were above 0.24 ppm for less than one-half of one percent of the total hours monitored during this same five-year period.[3]

And Los Angeles is not an isolated example. Chicago, Atlanta, Portsmouth, and most other nonattainment areas each exceeded the standard an average of less than one percent of the total monitored hours. Of course, "all monitor hours" include nighttime hours and winter months, which typically do not present an ozone problem.

The point is that equating the term "nonattainment" with unhealthful is an oversimplification. Without knowledge of the health effects of ozone and a more realistic perspective of the numbers of individuals exposed to elevated ozone levels and the frequency and duration of that exposure, the general public is presented with a distorted view of America's smog difficulties.

Ozone Health Effects

Health effects of ozone have been studied extensively during the past two decades. Both short-term (acute) effects and long-term (chronic) effects have been examined, but by far the greatest amount of research has been on acute responses to elevated levels of ozone.

Medical research has used two basic types of studies—epidemiological and clinical. Epidemiological research on pollutants usually examines hospital and doctor records to determine if there is a relationship between a given health effect and some environmental factor such as elevated ozone levels. A variety of confounding variables and lack of information on individuals' exposures to ozone have typically made strong results difficult to obtain from these studies.

Clinical studies, on the other hand, are conducted in controlled settings where the environment experienced by the subject is under the researchers' control. But these studies have been criticized because they typically use small numbers of subjects who have not been randomly selected.

Not all clinical studies are small-scale, however. For example, McDonnell *et al.* studied a group of 135 healthy, young, adult men who lived in low-pollution areas.[4] The test group was segmented into six subgroups, and each subgroup was exposed to one of six concentrations of ozone (0.12, 0.18, 0.24, 0.30, and 0.40 ppm) while exercising very heavily, but intermittently, for two hours. Small changes in breathing capacity were observed at 0.12 and 0.18 ppm. Only at the higher levels (0.24 and above) were average lung function losses greater than 10 percent. Although average lung function losses were mild even at the 0.18 level (50 percent above standard), individual responses varied greatly—and some subjects experienced severe decreases in lung function while others experienced no decreases whatsoever.

Summarizing the results of the myriad of clinical studies that have been conducted is beyond the scope of this essay. In general, while the studies show adverse health effects at high ozone concentrations, there are a few points that need clarification.

Ozone's effects on pulmonary function appear to involve an attenuation response, albeit a *temporary* one. Results from clinical studies show that, with repeated exposure to ozone, reductions in pulmonary function are greatest on the second day. On each succeeding day, the reductions are less than the day before. (Attenuation of a symptomatic response at a given ozone concentration does not reduce response to higher levels, however.[5]) Within three to seven days following the cessation of a sequence of repeated daily exposures, pulmonary function apparently returns to that experienced prior to exposure.[6]

Many of the concerns about ozone's effects on persons suffering from lung disease also lack substantiation. According to available evidence, people with pre-existing lung disease respond similarly to normal, healthy subjects when exposed to moderate ozone concentrations and exercise levels. This does not imply, however, that persons with already reduced lung function are not more at risk when experiencing the same *incremental* loss of breathing capacity as healthy subjects.

Short-term changes in lung function and increased respiratory symptoms are especially affected by the frequency and depth of breathing, which increase as exercise work-load increases. The EPA has pooled a variety of controlled human exposure and field studies to estimate the relationship between reduced breathing capacity and various combinations of exercise levels and ozone concentrations. Figure 3.1 illustrates these findings for healthy, adult subjects (18 to 45 years old) after one to three hours of exposure.

These overall results show less than a 10 percent loss in lung function at ozone levels more than four times the current standard during light exercise. Even during very heavy exercise, pulmonary function is typically reduced by less than 10 percent at 0.24 ppm—twice the Clean Air Act standard.

The EPA is quick to point out, however, that losses in lung function can occur with very heavy exercise in some healthy adults at 0.15 to 0.16 parts per million, and adverse effects have been shown at levels as low as 0.12 ppm. Children also may show decreases in lung function at levels as low as 0.12 ppm with heavy exercise.[7] Therefore, approximately 5 to 20 percent of the populations studied in these clinical tests have been dubbed "responders," showing a greater responsiveness than average subjects to the same conditions.

Despite the extensive clinical and epidemiological studies done thus far, the book on ozone health effects is far from closed. Studies of ozone's chronic effects on animals, for example, continue to raise concerns among medical researchers. Some ani-

FIGURE 3.1
Lung Function Decrements for Varying Ozone and Exercise Levels [a]

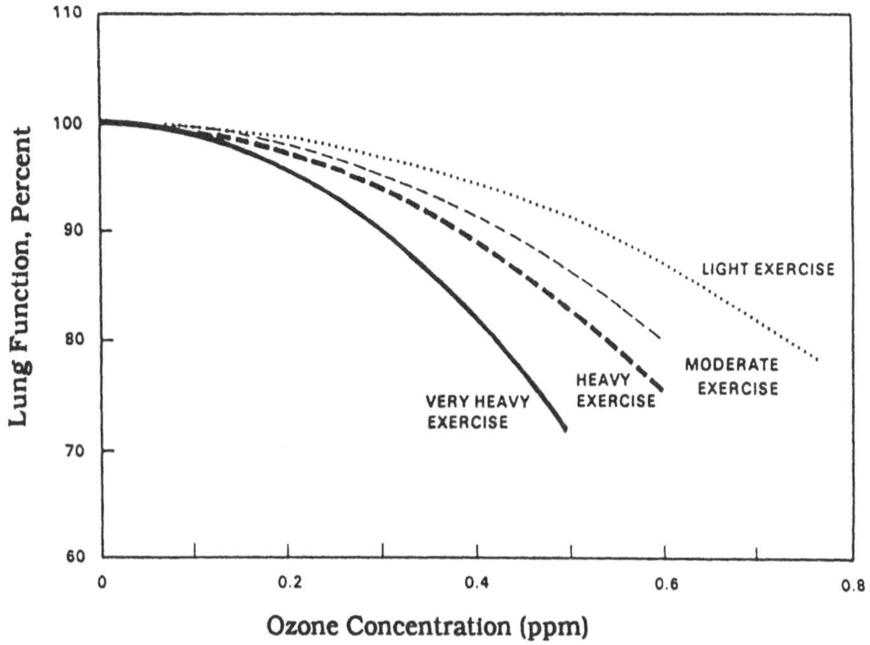

[a] Lung function is measured as the volume of air a subject can force from his/her lungs in one second.

Source: *Air Quality Criteria for Ozone and Other Photochemical Oxidants*, Vol. I (Washington, D.C.: U.S. Environmental Protection Agency, Environmental Criteria and Assessment Office, August 1986), p. 1-158.

mal research indicates that ozone affects the lung's ability to resist bacterial and viral infections and accelerates the lung's aging process.[8] Extrapolating these findings to humans is highly problematic, however.

In summary, elevated concentrations of ozone clearly produce adverse health effects. These effects are greatly influenced by levels of exercise, thus suggesting that significant numbers of at-risk individuals may be able to reduce the effects of smog by altering their behavior. Further, the demonstrated effects have largely been short-term (acute) effects of a relatively mild and reversible nature. On the other hand, there is a clear need for further medical research on possible chronic effects of ozone.

Causes of Ozone Pollution

Scientists also have been studying extensively the causes of ozone formation. This major component of photochemical smog is not emitted directly into the air, but is formed through complex chemical reactions between emissions of volatile organic compounds (VOCs)—primarily hydrocarbons—and nitrogen oxides (NO_x) in the presence of sunlight and oxygen. Both hydrocarbons and nitrogen oxides are emitted by transportation, residential, and industrial sources.

Ozone production follows several distinct patterns. Since the formation of ozone at the earth's surface requires sunlight, concentrations are minimal around sunrise (near zero in most urban areas), rise to maximum levels in the early afternoon, and fall to minimal levels again at night. Ozone also follows a seasonal pattern. During the late spring and summer, when sunlight is most intense and stagnant meteorological conditions are present, conditions are best for the formation and accumulation of ozone in the United States.

Some factors contributing to elevated ozone levels are beyond human control. One component of ozone formation (VOCs) occurs naturally due to emissions from trees and plants. While some investigators believe that these naturally occurring sources are the dominant contributors of VOCs, other researchers contend that at least two-thirds of these emissions result from man-made sources.[9]

Because ozone moves with air masses, ozone levels can be higher in suburban or rural areas than in urban areas. Moreover, elevated levels of ozone can persist longer in outlying areas because of the absence of nitrogen oxide for chemical "scavenging," breaking down the ozone molecule into harmless oxygen. Under some circumstances, reductions in nitrogen oxide concentrations can reduce its scavenger role and actually increase the formation of smog. The dual role of NO_x as both a contributor to and a scavenger of ozone is one of the factors that makes it so difficult to deal with ozone problems in an across-the-board fashion.

In summary, the combination of weather conditions and natural and man-made emissions of pollutants that form ozone make the objective of decreasing peak ozone levels problematic. A regulatory approach that does not take specific local circumstances into account is likely to produce disappointing results.

The Costs and Benefits of Reducing Ozone Levels

There are no "silver bullets" that can bring into compliance all of the 83 urban areas that did not meet the 0.120 ppm ozone standard during the 1986-1988 period. Several methods required by laws which predate the 1990 CAA amendments can reduce 1987 VOC emission levels by a total of 18 percent by 2005. By 1995, these measures alone should reduce the out-of-compliance areas to 58; however, without added controls, population growth and increased automobile usage could increase that number to 72 by 2005.[10]

The Bush administration's clean air proposal identified ten new measures that could reduce VOCs by an additional 27 to 30 percent. The White House projected that these added measures would bring all but 20 cities into compliance by 2005. These twenty areas that would continue to be out of compliance have a current population of 36 million.

Overall Benefit Estimates

The Clean Air Act does not require that decisions relating to control measures be based on cost/benefit or cost-effectiveness analysis. As a result, there are few estimates of the costs or benefits of achieving improvements in urban smog. Nonetheless, recent reports do provide some rough but useful estimates of aggregate benefits and costs.

Health Benefits. One comprehensive study of acute human health (and agricultural) benefits, completed for the Office of Technology Assessment (OTA) by Alan Krupnick and Raymond Kopp of Resources for the Future, provides benefit estimates associated with various decreases in ambient ozone concentrations.[11] The Krupnick/Kopp study uses a willingness-to-pay method of benefit estimation, and focuses upon reductions in acute health effects. People are surveyed to find out how much they would be "willing to pay" to avoid a day (or part of a day) of a health effect or group of effects, and the results are extrapolated for the whole exposed population.

Krupnick and Kopp present seven scenarios and calculate benefits for each of them. The two researchers found that their estimates of health benefits could differ by as much as a factor of one hundred. Depending upon the assumptions, the benefits of reaching the current ozone standard of 0.120 ppm range from $51 million to $4.7 billion a year (in 1984 dollars), which is quite a range. Assuming benefits are derived only by heavily exercising individuals, the payoff from reaching the 0.120 ppm standard ranges from $51 million to $360 million, while, if all exercising individuals are presumed to receive benefits proportional to exercise levels, the benefits range from $667 million to $4.7 billion.

Of greatest interest for comparison of benefits and costs is Krupnick and Kopp's analysis of an across-the-board 35 percent reduction in VOCs. Assuming that all exercising individuals receive benefits, a 35 percent reduction in emissions would produce health benefits ranging from $248 million to $1.7 billion with a midpoint of $684 million. Taking the high end of this range and adjusting benefits to 1988 dollars, acute health benefits of reducing VOC emissions would amount to $500 a ton.

Other Benefits. Reducing ozone levels also benefits crops and forests. A good deal of research has gone into this area, and benefit estimates are available for agricultural crops. But benefit estimates for human health and agricultural products are not calculated on a comparable basis. Furthermore, control strategies to reduce VOC emissions in urban areas may produce few crop benefits in rural areas and vice versa.

Other benefits from reduced concentrations of ozone (such as increased property values) might also be postulated, but the primary benefits, and thus those of most concern, are improvements in public health. Likewise, the costs associated with achieving more healthful air in cities and towns is of more interest than possible costs of improving rural air quality.

White House and OTA Estimates of Aggregate Costs

Aggregate cost projections of meeting the national air quality standard for ozone are derived by adding up piecemeal estimates of costs for the various proposed requirements. For example, President Bush's initial proposal, designed to bring all but 20 cities into ozone compliance by the year 2005, was purported to cost $3-4 billion a year.[12] The ten new steps included in this plan were estimated to reduce emissions by 27 to 30 percent from 1987 levels.

One of the most comprehensive estimates of aggregate costs for new methods of reducing ozone levels is contained in a July 1989 Office of Technology Assessment report, "Catching Our Breath." OTA projects total reductions of about 34 percent by 1994 at a cost for all nonattainment areas between $4.2 and $7.1 billion per year. Costs would rise to between $6.6 and $10 billion annually by 2004.[13]

Our own estimates, based upon EPA and OTA data, indicate that it would be possible to reduce emissions by 40 percent by the year 2005 at an estimated total nationwide cost of approximately $8.5 billion a year (in 1986 dollars).[14] Theoretically, the combination of control measures used to achieve these results would bring all but 38 areas into attainment by the year 2005.

Comparisons of Costs and Benefits

Table 3.2 shows benefit-to-cost ratios assuming the most inclusive health benefit estimate from the Krupnick and Kopp study compared to White House, OTA, and our own study's projections of the cost-effectiveness of control techniques. These comparisons are very rough, however, and should be considered only as "indicators."

The Bush cost estimates and effectiveness projections suggest that the average cost to remove a ton of VOCs is in the $1,200-1,500 range. The ratio of health benefits to abatement costs using White House projections thus ranges from 0.3 to 0.4. The benefit-to-cost ratios are slightly worse using OTA estimates. Costs per ton range from $1,700-2,600 and the benefit-to-cost ratio ranges from 0.2 to 0.3. We estimate average nationwide costs for reducing VOCs to be $2,100 a ton, for a benefit-to-cost ratio of roughly 0.2.

Keep in mind, however, that this analysis does not include agricultural benefits or any additional costs to reduce rural ozone concentrations. In addition, other non-health benefits were not included in these estimates.

To be certain, no definitive conclusion should be drawn from such crude estimates, but benefit-to-cost ratios substantially less than 1.0 raise a warning flag. In general, optimum regulatory levels would occur where marginal benefits and marginal costs are equal—a point that typically occurs at a level of regulatory stringency less than the point where total benefits and costs are

TABLE 3.2
Comparing Acute Health Benefits to Costs
(in 1988 dollars)

	Health Benefits[a]	VOC Reduction in Areas of Nonattainment[b]	Average $Benefit Per Ton
Krupnick and Kopp	$1.9 billion	35% (3.8 million metric tons)	$500
	Abatement Costs	**VOC Reduction in Areas of Nonattainment[b]**	**Average $Cost/Ton**
White House Proposal	$3-4 billion	30% (2.6 mmt)	$1,200-1,500
OTA	$6.6-10 billion	35% (3.8 mmt)	$1,700-2,600
Center for the Study of American Business	$9 billion	40% (4.3 mmt)	$2,100

Benefit-to-Cost Ratio	
Krupnick and Kopp/White House	.33 - .42
Krupnick and Kopp/OTA	.19 - .27
Krupnick and Kopp/CSAB	.24

[a] The highest estimate of benefits from a 35 percent rollback of VOCs is $1.7 billion in 1984 dollars, the high estimate from clinical studies.

[b] All VOC reductions are for a baseline of 1985 levels (11 million metric tons) except for the White House figures. The White House assumed 1987 baseline levels. A figure of 8.6 million metric tons for 1987 VOC emissions for non-attainment areas was derived by taking nationwide emissions of 19.6 mmt from Table 3.5 of the *National Air Quality and Emissions Trends Report, 1987*, EPA, and multiplying it by 44 percent. The 44 percent factor was derived by dividing non-attainment by total emissions figures contained in Tables 6-1 and 6-4 in *Catching Our Breath: Next Steps for Reducing Urban Ozone* (Washington, D.C.: U.S. Office of Technology Assessment, July 1989).

Source: Authors' calculations.

equal. In and of themselves, benefit/cost ratios as low as these indicate that inefficient regulation is occurring—consumer and taxpayer dollars are likely being misallocated.

Effectiveness of Control Measures

Aggregate cost figures do not provide sufficient information for making decisions about particular control measures, however. A brief analysis of the cost for each ton of VOCs removed for several specific measures follows.

Cost-effectiveness evaluations are not available for many of the control procedures being proposed for reducing ozone levels. Nonetheless, EPA and the Office of Technology Assessment have gathered cost-effectiveness estimates for a variety of control measures that apply on a nationwide basis. Using this data we have analyzed the potential emission reductions, the cost per ton, and the total cost of reducing VOC emissions for a variety of measures: (1) applying reasonably available control technology to point and area sources; (2) requiring an enhanced inspection and maintenance program for vehicles; (3) instituting transportation control measures; (4) reducing fuel volatility; (5) requiring service stations to install Stage II fuel recovery systems; and (6) mandating onboard fuel recovery systems for autos.

Figure 3.2 illustrates the percentage of 1985 baseline VOC emissions potentially reduced by each of these control strategies in the years 1995, 2000, and 2005. These estimates are nationwide averages for all nonattainment areas; actual reductions almost certainly vary across cities. The reductions in baseline emissions shown in this figure are not net emissions, however. Population growth and increased automobile traffic will add to the emissions inventory and thus partially offset benefits derived from added controls.

Reasonably Available Control Technologies (RACT). The category of controls producing the largest reduction in VOC emissions, is called reasonably available control technologies. This category consists of a variety of control measures, each applicable to a particular source of VOCs (*e.g.*, dry-cleaning, petroleum refining, chemical manufacturing, paper coating, etc.). In all, OTA and EPA supplied information on control technologies for 40 specific source categories that together emit a substantial portion of controllable VOCs. From this data, it appears that VOC emissions can be reduced by about 14 percent (from 1985 baseline levels) by applying reasonably available control technologies.

FIGURE 3.2
Percent Reduction for Selected Controls
from 1985 Baseline VOCs
(for 1995, 2000, 2005)

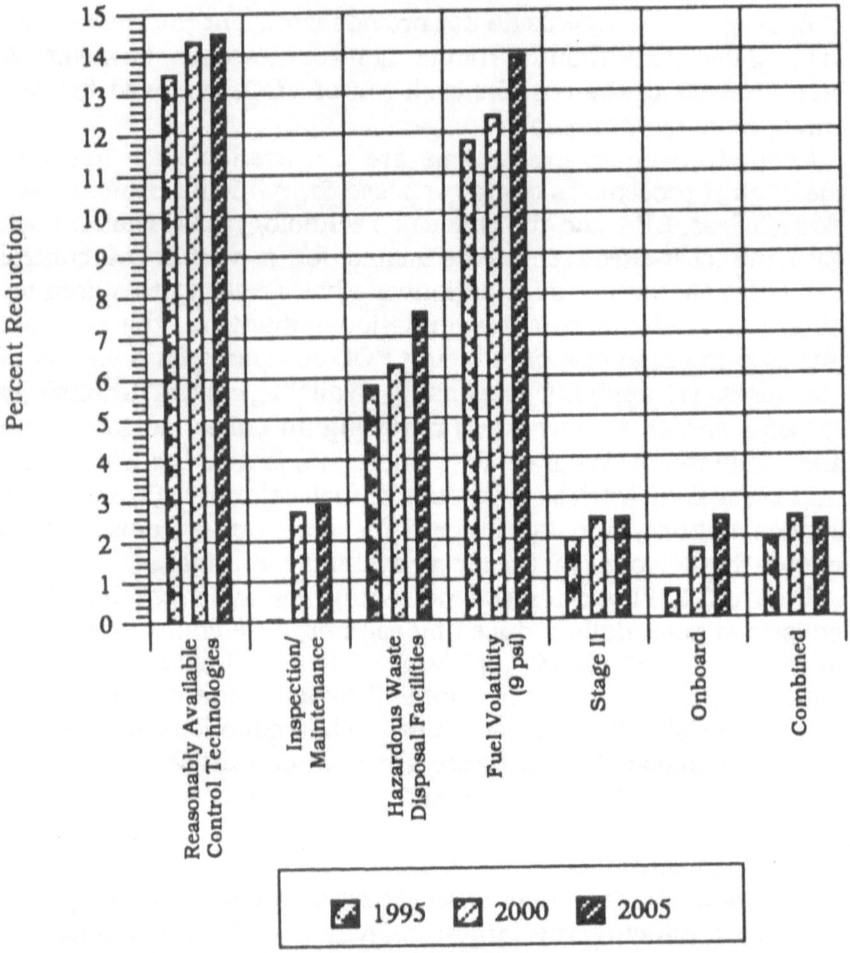

Source: Author's calculations for fuel volatility are from Tables 3-21, 3-22, 6-3, 6-4, 6-5, and 6-6 in *Draft Regulatory Impact Analysis: Control of Gasoline Volatility and Evaporative Hydrocarbon Emissions from New Motor Vehicles* (Washington, D.C.: U.S. Environmental Protection Agency, July 1987). Calculations for other control methods from data in *Cost Assessment of Alternative National Ambient Air Quality Standards for Ozone*, Draft Report, October 1987, Alliance Technologies Corporation. Prepared for U.S. EPA Office of Air Quality Planning and Standards.

Figure 3.3 illustrates the average dollar cost per ton of VOCs removed (unit cost) for each of the selected emission control methods. Clearly, reducing fuel volatility is the most cost-effective, costing about $250 to remove a ton of VOCs. Applying all known "reasonably available control technologies" to stationary sources involves high costs for each ton of VOCs removed, about $4,400 a ton.

Figure 3.4 presents a dilemma: requiring the additional installation of reasonably available control technologies, while quite expensive as a whole, is also capable of reducing the largest percentage of emissions. Not all "available technologies" can be considered "reasonable" when imposed on small sources, however. For instance, incineration of VOCs at all small surface-coating plants is estimated to cost just over $7,700 a ton. Similarly, controls at small dry cleaners average $3,600 a ton while these same controls at large dry cleaners (at least 100 tons of VOC emissions a year) cost an average of only $230 for each ton of VOCs removed.

Unfortunately, the emissions reduced by these very expensive controls make up a significant portion of the expected VOC reductions obtainable through applying reasonably available control technologies more widely. If control of these small sources is not assumed, VOC reductions of approximately 8 percent of the 1985 baseline emissions can be achieved by 2000 as a result of applying the remaining technologies. Correspondingly, the cost to eliminate a ton of VOC emissions using reasonably available control technologies is reduced by 40 percent from $4,400 a ton to $2,800 a ton.

Fuel Volatility and Vehicle Emissions Certification. Another method capable of significantly reducing VOC emissions is lowering fuel volatility. The 1990 CAA amendments require that by 1992 the volatility of gasoline in nonattainment areas during the ozone season not exceed a Reid Vapor Pressure of 9.0 pounds per square inch.

The effects of this control measure are twofold. First, lowering fuel volatility significantly reduces "running losses"—evaporative emissions that occur while a vehicle is in operation. Second, the more volatile the fuel, the less effective the auto emission control systems, resulting in actual hydrocarbon emissions that are higher than standards dictate. Taking both of these effects into account, lowering fuel volatility to the current level used in testing pollution equipment—9.0 psi—would reduce VOC emissions by 11.5 percent in 1995 and by 13.5 percent in 2005.

FIGURE 3.3
Unit Cost Estimates for Selected Controls
(for 1995, 2000, 2005)

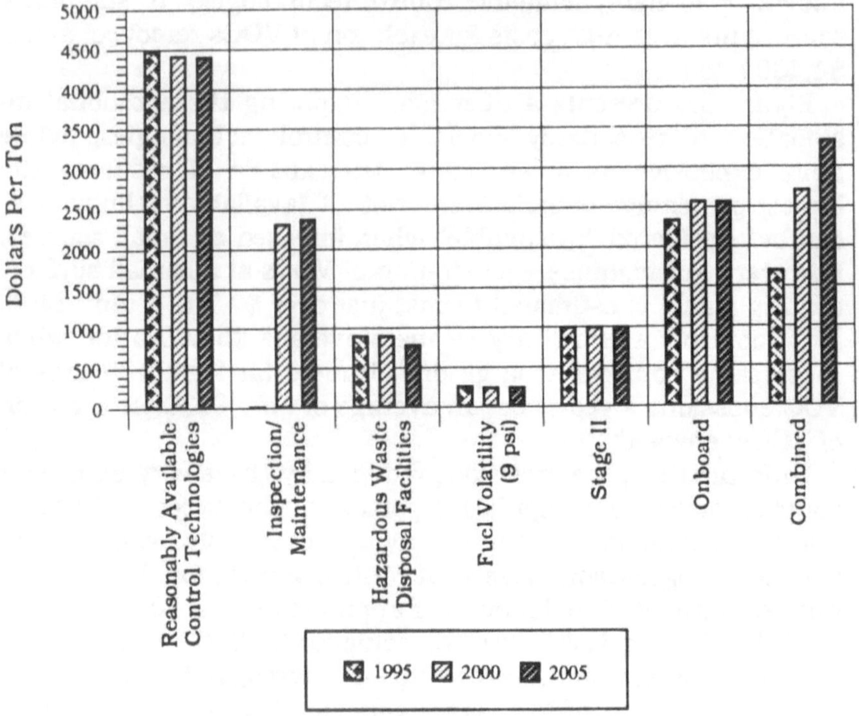

Source: See source line for Figure 3.2.

Stage II and Onboard. Refueling emissions at service stations can be controlled by either of two vapor-recovery technologies: (1) new cars can be required to be equipped with "onboard" controls or (2) service stations can be required to install "Stage II" equipment on each of their gasoline pumps. Since practicality dictates that only new cars be equipped with the onboard device, *at least* ten years would pass before full effectiveness of the onboard option would occur. (See Figure 3.2.) Further, because nationwide, rather than regional, implementation is a more feasible strategy, adoption of the onboard technology may cause *all*

FIGURE 3.4
Unit Cost Estimates and Possible VOC Reduction for Selected Controls by 2005

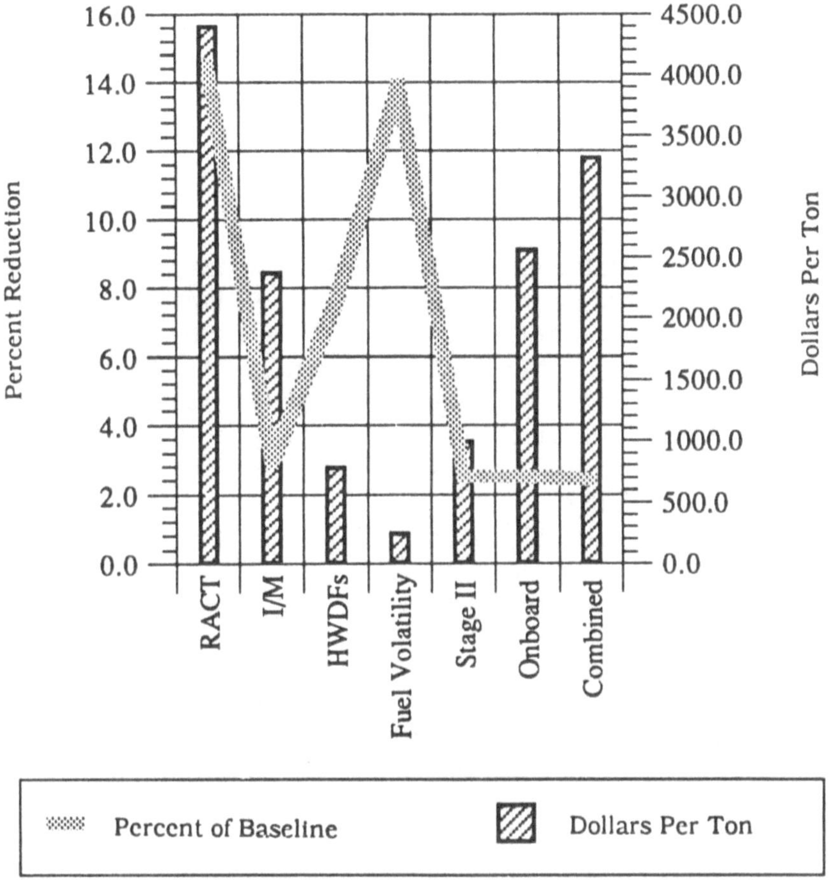

Note: RACT = Reasonably Available Control Technologies; I/M = Inspection/Maintenance; and HWDFs = Hazardous Waste Disposal Facilities.

Source: See source line for Figure 3.2.

the driving public to bear the costs uniformly, while nonattainment areas receive disproportionately large benefits.

Both strategies for controlling vehicle refueling emissions—Stage II and onboard—are capable of attaining basically the same VOC reductions in nonattainment areas by the year

2005, about 2.5 percent. But the Stage II controls achieve much larger immediate results and are more than twice as cost-effective as onboard controls, costing about $1,000 a ton versus $2,600 a ton of nonattainment VOCs removed using the onboard device.

The least cost-effective approach of all, however, is to require both Stage II and onboard equipment. The combined strategies produce no added benefits to nonattainment areas, but the costs of implementation increase dramatically. This approach is a case of getting nothing for something.

Unfortunately, this is the strategy adopted in the CAA amendments. Specifically, any area classified as at least a "Moderate" nonattainment area, (areas with design values of 0.138 and above) must implement Stage II technologies at most service stations beginning 6 months to 2 years after enactment. (President Bush signed the 1990 CAA amendments on November 15.) Further, since the EPA administrator may reclassify areas whose design values are within 5 percent of the next classification, the areas that could be affected include those with design values as low as 0.131.

Pending the results of a safety study, vehicle manufacturers must begin equipping light-duty vehicles with onboard systems as early as the fourth model year after the EPA administrator promulgates the standards. (The standards must be put into effect within a year after enactment.) By the sixth model year, all of these vehicles must be equipped with the onboard device. Once the onboard implementation is widespread, the installation of Stage II systems may be discontinued at the EPA administrator's discretion.

The reasoning behind this decision is based on the belief that Stage II systems will cause undue hardship on small service stations that would need to increase their per gallon charges for gasoline more than their larger competitors. Because of this hardship, small stations are excused from Stage II implementation, reducing the measure's potential for controlling VOCs. On the other hand, because onboard devices will not achieve immediate results, Stage II is used to fill the void.

Much of the unnecessary cost from this approach could be avoided by a simple tax/subsidy plan designed to remove the competitive problems faced by small service stations. All service stations in nonattainment and transport areas could be fitted with the Stage II devices, and the residents of the included areas could pay a tax—from $1.05 to $2.70 a person—to cover the implementation costs. Costs for nationwide onboard implementation may be $14.00 a person regardless of whether the individual lives in a non-attainment area.[15] It appears that taxpayers' dollars are being wasted and could be better used to produce real improvements in air quality.

Other Measures. Other control methods analyzed include transportation control measures—carpooling and alternative working hours, expanded vehicle inspection and maintenance, and controls on hazardous waste disposal facilities.

Figure 3.3 shows that removing a ton of VOCs using an expanded inspection and maintenance program is about nine times as costly as using fuel volatility controls. In addition, lowering fuel volatility can reduce emissions five times as much overall.

As Figure 3.2 shows, transportation control measures are not expected to greatly reduce ozone levels nationwide—less than one percent. But there are certain areas of the country where such controls may be more beneficial—specifically, the South Coast Basin in California (which includes Los Angeles).

In September 1988, the South Coast Air Quality Management District proposed a three-tier, twenty-year plan to achieve ozone attainment. The transportation control measures (including management of growth) included in this plan provide a best-case scenario of possible VOC reductions from these controls.

By the year 2010, the South Coast Basin expects cars, trucks, and other mobile sources to contribute 39 percent of total VOC emissions. Transportation control measures proposed include: (1) requirements to reduce work trips by at least 10 percent; (2) increases in carpooling and ridesharing resulting from preferential parking and financial incentives for participants, ride matching services, park-and-go lots, high occupancy vehicle lanes, and a cap on the number of parking spaces available; (3) transit improvements, such as increased bus fleet and express services; (4) truck rerouting; (5) traffic flow improvements, including ramp metering and synchronized signals; (6) growth management; and (7) electrification of some vehicles and rail.

All of these controls are predicted to result in a 12 percent decrease in 2010 baseline VOC levels for the South Coast, almost a one-third reduction of emissions from mobile sources.

The cost, however, is virtually unmeasurable. Lifestyle changes, particularly growth management measures that account for one third of the anticipated reductions, are especially difficult to cost out. Moreover, as the OTA cautioned in its July 1989 report, "Involuntary transportation control measures have proven politically infeasible and voluntary ones difficult to sustain."[16]

Temporal Controls. One group of control measures currently given short shrift is temporal controls—measures, such as staggered work hours and reduced commuter traffic that alter individual and business behavior during critical time periods. With the exception of "extreme" nonattainment areas, the Clean Air Act prohibits including temporal controls in state implementation plans, requiring that all areas attempt to meet the standard using controls that are permanent. As a result, the cost savings possible from using temporal controls to reduce ozone levels have not been investigated. One study of continuous versus intermittent abatement techniques for controlling sulfur dioxide in Tennessee found that the cost of meeting the sulfur dioxide standard using temporal measures would be one-fifth the cost of constant controls.[17]

Moreover, health benefits from time-varying controls can be substantial in cases where continuous controls are unable to prevent high ozone peaks. While there is much debate about possible threshold levels for ozone effects, it is clear that higher levels have more pronounced effects on larger numbers of individuals. Thus, flattening out peak ozone concentrations with temporal controls may produce greater health benefits than continual measures that reduce VOC emissions by a constant amount at all times.

Legislative Recap

The original Clean Air Act proposal from the Bush administration emphasized using market incentives and allowing regulatory flexibility in order to tackle air pollution problems in a more efficient way. A key feature of that proposal was maintained in the final legislation—establishing categories of nonattainment based upon an area's design value. By grouping nonattainment areas into five different classes—marginal (0.121 to 0.138 ppm), moderate (0.138 to 0.160 ppm), serious (0.160 to 0.180 ppm), severe (0.180 to 0.280 ppm), and extreme (0.280 ppm and above)—the "cure" can better be prescribed to fit the "disease." The worst cases of smog require the most stringent methods of reducing ozone and these areas are therefore given more time to meet the standard. A brief recap of the major features of the

1990 Clean Air Act amendments that relate to ozone seems in order.

Ozone Requirements

All nonattainment areas. There are two provisions required of all nonattainment areas, and both of these relate to mobile sources. As already discussed, nationwide implementation of onboard systems will be required pending review of the safety of this equipment. In addition, all gasoline sold in nonattainment and transport areas will be required to meet ozone season volatility regulations. These regulations specify that the Reid Vapor Pressure of gasoline not exceed 9.0 pounds per square inch (psi) during a specified ozone season, which may vary depending upon the area.

Interestingly, because alcohol raises gasoline vapor pressure, legislators have stated that alcohol blends should be given a 1.0 psi waiver from the volatility requirements. The CAA amendments do this, but only for ethanol blends—not any methanol blends. This seems a rather obvious concession to the farm lobby and ethanol producers that is hard to justify from an environmental standpoint.

Marginal areas (0.121 to 0.138 ppm). Very little in the way of specific regional controls is required for marginal areas because many existing controls, as well as those controls required nationwide, are expected to bring these areas into attainment within three years. The CAA amendments require these areas to revise both vehicle inspection and maintenance programs and applications of RACT to make sure the areas will meet requirements that were already prescribed before enactment. Further, these areas must revise their permit programs to bring them up to date with any new interpretations made by the EPA administrator. New sources are required to meet offset ratios of 1.1 to 1, *i.e.*, for every 10 tons of new emissions needed by a new facility, 11 tons must be eliminated elsewhere in the same nonattainment area.

Moderate areas (0.138 to 0.160 ppm). Areas classified as moderate must fulfill all of the requirements for marginal areas and must comply with a number of additional measures. RACT is required for all major sources, which are defined as those sources with an annual emission of at least 50 tons of VOCs, and for all sources covered by existing control technology guidelines. The Stage II vapor recovery system must be installed in all service stations selling more than 10,000 gallons of gasoline per month (50,000 gallons per month for independent small business marketers): within six months for stations commencing construction after enactment, within one year for those stations selling at least 100,000 gallons per month, and within two years for all other stations. Also, a vehicle inspection and maintenance program must be implemented in these areas, regardless of whether such a requirement existed previously. Any offset requirements pertaining to new sources in moderate areas must be at least 1.15 to 1.

Serious areas (0.160 to 0.180 ppm). Nonattainment areas classified as serious are required to implement all the provisions for moderate areas and numerous additional measures. (Major sources are those sources emitting at least 50 tons of VOCs annually.) Enhanced vehicle inspection and maintenance, which includes annual testing, on-road testing devices, fewer vehicle waivers, and enforcement through the denial of registration, is required.

In addition to implementing an enhanced monitoring program, these areas must demonstrate that their revised plan will meet reasonable further progress, *i.e.*, that the area can reduce baseline emissions by an average of at least 3 percent a year (over a rolling three-year period beginning six years after enactment). These reductions are referred to as "milestones" and cannot include emissions abatement that results from nationally implemented programs, such as reduced fuel volatility or further reductions in tailpipe emissions, or from corrections to implementation plans or vehicle inspection and maintenance plans.

Recognizing that NO_x may also play an important role in ozone formation, areas in this classification may demonstrate reasonable further progress by reducing a combination of VOCs and NO_x as long as it produces a reduction in ozone concentrations equivalent to those that would be achieved from the mandated VOC reductions. This represents a valuable realization of the complex local chemistry of smog discussed in an earlier section of this essay.

Unless substitute measures are approved, a clean fuels program shall be implemented within 42 months. Such a program phases "clean vehicles"—those running on natural gas, reformulated gasoline, 85 percent alcohol mixtures, or electricity—into fleets operating in serious nonattainment areas. Generally, by the year 2000, 70 percent of the light-duty fleet vehicles must be clean vehicles. The emissions allowed for these vehicles are basically one-half of those allowed for conventional vehicles under the CAA amendments. Beginning six years after enactment, serious areas must show that aggregate vehicle mileage, emissions, and congestion are consistent with the areas' attainment plans. If not, the area must revise its implementation plan to include transportation control measures.

Any offset requirements pertaining to sources in this area must be at least 1.2 to 1. If a source wishes to expand, it can avoid specific control technology required for modifications if the source can provide an internal offset of at least 1.3 to 1. Otherwise, if the source emits less than 100 tons of VOCs per year, best available control technology (BACT) will apply, and if the source emits 100 tons or more per year, lowest achievable emissions rate will apply.

Severe areas (0.180 to 0.280 ppm). Added measures over and above those specified for serious areas are required for severe areas. (Major sources are redefined as those emitting annual VOCs of at least 25 tons.) Offset requirements pertaining to sources in this area must be at least 1.3 to 1 unless the entire state requires major sources to implement BACT. In this case, the offset ratio is reduced to 1.2 to 1.

Additional transportation control measures are required for severe areas. These measures include, but are not limited to, a reduction in work-related trips and miles traveled and employer-motivated increases in vehicle occupancy (of at least 25 percent) during peak traveling times.

Extreme areas (0.280 ppm and above). Extreme areas are required to implement all of the provisions for severe areas as well as the following additional provisions. (Major sources are redefined to include all sources emitting at least 10 tons of VOCs annually.) Any offset requirements pertaining to sources in this area must be at least 1.5 to 1, unless the state requires all major sources in the nonattainment area to use BACT. In this case, the required offset ratio is 1.2 to 1. As described for serious areas, a source may avoid being considered modified if it makes internal offsets of at least 1.3 to 1. This does not apply if the source is required to make these changes to fulfill other requirements of the CAA amendments.

Within eight years after enactment, new, modified, and existing electric utility, industrial, and commercial boilers that emit more than 25 tons per year of NO_x will be required to burn low-polluting fuels as the primary fuel and to use advanced controls for reduction of NO_x. These provisions apply only where the

EPA administrator deems them desirable. Also, the administrator may approve provisions for implementation that include new technologies whose development is anticipated.

Penalties for Non-Compliance

If serious or severe areas do not meet their milestones, three options are possible: (1) the areas can be reclassified to the next higher classification; (2) additional measures can be implemented so that the areas meet the next milestone; or (3) an economic incentives program can be implemented. The latter option allows for a system of emissions fees, or marketable permits, a tax on the sale or manufacture of products whose use contributes to ozone formation, or any combination of these. Such a system may also include incentives to lower vehicle miles traveled and emissions produced.

If an extreme area does not meet its milestones, the state must submit a revised program that includes the economic incentives program described above.

If severe or extreme areas fail to attain the air quality standard by the applicable deadline, each major source in the affected area must pay an annual fee for its emissions until the area reaches attainment status. The fee is calculated as $5,000 (1989 dollars) for each ton of VOCs emitted by the source in excess of 80 percent of the "baseline" amount. The baseline amount is the lower of the actual VOC emissions or of the permitted VOC emissions for that particular source. This fee does not apply to areas with a population under 200,000 who can also prove that they were not in attainment because of ozone or precursor transport from other areas.

Conclusion

Fundamentally, the problem with the new Clean Air Act is that it falls short of the true reform needed to improve air quality in the most cost-effective manner. Our estimates of the ratio of benefits to costs, ranging from 0.2 to 0.4 for a 30-40 percent reduction in ozone-producing emissions, are one indication of inefficiency.

What basic changes in the Clean Air Act would constitute fundamental reform of this flagship environmental law? First of all, rather than require that air quality standards provide an adequate margin of safety against *any* adverse health effects, the Clean Air Act should have as its basic objective "to protect the public against *unreasonable* risk of *significant* adverse health effects." Public health would be unaffected by this change, but achieving this goal would cost less, thereby reducing the adverse economic effects of ozone regulations.

Former EPA official Milton Russell recently concluded in his 1988 study of ozone for Oak Ridge National Laboratory that "the way the law [Clean Air Act] is now written, it is almost as if a cancer were equivalent to a cold, one expected cancer were indistinguishable from an epidemic and as much social disruption . . . and economic cost were to be imposed to avoid the one as to avoid the other."[18] The revised Clean Air Act does nothing to change this assessment. We can afford to protect ourselves from unreasonable exposure to unhealthful air pollutants, but it is impossible to create the zero-risk environment implicitly promised in the Clean Air Act.

Another basic problem with ozone provisions in the Act is that the definition of "nonattainment" does not reflect the true severity of the smog problem. An area is labeled nonattainment if the fourth highest daily monitor reading during the most recent three-year period registers an ozone concentration above 0.120 parts per million for more than one hour. Given that short-term health effects are reversible and depend on exercise levels (also on outdoor as opposed to indoor exposure), such a peak-period, geographically isolated definition seems unwarranted.

While the amended 1990 Clean Air Act takes a step in the right direction by defining categories of nonattainment and varying the prescriptions accordingly, it continues to rely on this faulty definition of nonattainment. Furthermore, due to the measurement period (1987-1989) used to establish these classes, the aberrant weather conditions experienced in the United States in 1988 will have an inordinate effect on the costs that individual cities will incur in their efforts to battle smog.

It would not be difficult to define classes of nonattainment in a fashion that is more consistent with public health risk. Rather than use the highest ozone reading recorded at a single monitor, an average should be taken of a representative set of monitors in a given area to determine if an "exceedance" has occurred. Furthermore, classification of nonattainment areas should be based on average annual exceedances, *as well as* some measure of peak ozone concentration, such as the 90th or 95th percentile of daily peak one-hour readings over a rolling three-year period.

One remote provision in the new Act provides hope that a better measure of the seriousness of nonattainment may be forthcoming. Not later than three years after enactment, the EPA administrator must submit a study to Congress evaluating whether the methodology currently used by EPA for establishing a design value provides a reasonable indicator of ozone air quality.

Another important principle underplayed in the 1990 Clean Air Act amendments is to allow problems to be solved at the lowest level of government possible. The objective should be to reduce the federal role in specifying precise control measures to be used. States should be encouraged to use innovative approaches that fit local circumstances in their state implementation plans.

Because each nonattainment area is unique, there is a need for greater flexibility than can be achieved by specification of numerous federal mandates. The EPA's role should be more one of "evaluator" of state implementation plans rather than monitor of state compliance with federal proscriptions. In this regard, temporal controls should be allowed in demonstrating compliance in state implementation plans for all nonattainment classes.

Once again the 1990 amendments take a tentative step in this direction. Within one year after enactment, the EPA administrator must provide guidelines for the States to use in evaluating the cost-effectiveness of various control strategies.

America's resources are vast but finite. Our analysis of atmospheric ozone—its health effects and its complex chemical nature—and of the Clean Air Act's provisions for dealing with urban smog, makes it clear that these resources could be allocated more effectively than at present. The foregoing reforms would not "trade lives for dollars"; on the contrary, in practice they would result in a higher level of social benefits for Americans but with less economic disruption.

Notes

1. *National Air Quality and Emissions Trends Reports, 1987* (Washington, D.C.: U.S. Environmental Protection Agency, March 1989), p. 3.

2. U.S. Environmental Protection Agency, "EPA Lists Places Failing to Meet Ozone or Carbon Monoxide Standards," *Environmental News*, July 27, 1987, Table 1.

3. *Ozone Concentration Data* (Washington, D.C.: American Petroleum Institute, 1987).

4. W. F. McDonnell, D. H. Horstman, S. Abdul-Salaam and D. E. House, "Reproducibility of Individual Responses to Ozone Exposure," *American Review of Respiratory Disease*, 131:36, 1985.

5. *Review of the National Ambient Air Quality Standards for Ozone and Other Photochemical Oxidants* (Washington, D.C.: Office of Air Quality Planning and Standards, November 1987), p. VII-46.

6. *Air Quality Criteria for Ozone and Other Photochemical Oxidants*, Vol. V, (Washington, D.C.: U.S. Environmental Protection Agency, August 1986), pp. 10-47.

7. *Air Quality Criteria for Ozone and Other Photochemical Oxidants*, Vol. I, pp. 1-156.

8. *Air Pollution: Ozone Attainment Requires Long Term Solutions to Solve Complex Problems* (Washington, D.C.: U.S. General Accounting Office, January 1988), p. 19.

9. *Air Quality Criteria for Ozone and Other Photochemical Oxidants*, Vol. I, pp. 1-15.

10. *Fact Sheet: President Bush's Clean Air Plan* (Washington, D.C.: The White House, June 12, 1989), p. 7.

11. Alan J. Krupnick and Raymond J. Kopp, *The Health and Agricultural Benefits of Reductions in Ambient Ozone in the United States* (Washington, D.C.: Resources for the Future, Discussion Paper QE88-10, August 1988).

12. *Fact Sheet: President Bush's Clean Air Plan*, p. 8.

13. Ibid., p. 12.

14. Kenneth Chilton and Anne Sholtz, *Battling Smog: A Plan for Action* (St. Louis: Center for the Study of American Business, 1989), p. 31.

15. Anne Sholtz, *The Cost-Effectiveness of Proposed Measures for Controlling Refueling Emissions*, Working Paper No. 131, (St. Louis: Center for the Study of American Business, 1989), p. 6.

16. Ibid., p. 19.

17. Azriel Teller, "Air Pollution Abatement: Economic Rationality and Reality," in Roger Reville and Hans Landsberg, eds., *America's Changing Environment* (Boston: Beacon Press, 1970).

18. Milton Russell, *Ozone Pollution: The Hard Choices* (Oak Ridge, Tenn.: Oak Ridge National Laboratory, September 1988), p. 43.

PART TWO

Putting Environmental Risks in Perspective

4

Setting Goals for Environmental Safety

Lester Lave

The public is concerned, perhaps even to the point of paranoia, about a host of hazards associated with current production and lifestyles, including pesticide residues in food, living near toxic waste dumps or nuclear power plants, highway safety, and the safety of air travel, to name a few. Supermarkets in California and Massachusetts now compete by advertising that their produce is free from pesticides. A few years ago, many people threw away boxes of cake mix or cornbread mix out of fear that tiny residues of EDB (the pesticide ethylene dibromide) might cause cancer. In 1989 consumption of red apples, apple juice, and other apple products all but stopped because of fear of Alar residuals. During the same period, the importation of all fruit from Chile was suspended and consumers were urged to dispose of all fresh produce that could have come from Chile—all because two grapes were found to have been injected with small amounts of cyanide.

In these and many similar cases, the concerns could not and cannot be dismissed as irrelevant. Some theoretical models such as the "one-hit" theory predict that cancer could be caused by even the most minute exposure to a carcinogen.[1]

The primary focus of much of the public's health concern is on chemicals. There happens to be a widespread association of the increased incidence of cancer with greater use of synthetic substances.

Since 1945, the production and use of synthetic organic chemicals has increased from about one billion pounds a year to more than 300 billion pounds a year, as shown in Figure 4.1.[2] In addition to increases in quantity, there has been an increase in the number of chemicals produced: about 50,000 to 60,000 chemicals are produced and used in nontrivial amounts.[3] Of these chemicals, perhaps one percent have been subjected to reasonably thorough testing, and the vast majority are essentially untested.

No expert can assure the public that trace contaminants in food colors or in drinking water will not lead to cancer. No expert can be certain that a nuclear reactor will not have an incident that spills radioactive elements into the environment.

The world is, and always has been, a dangerous place. At the risk of seeming cavalier about the subject, we need to remember that eventually, every person dies, no matter how well loved, meritorious, or rich.

Until this century, the largest cause of death was infectious diseases, such as smallpox and tuberculosis. The bubonic plague almost depopulated cities in the 14th century, killing 25 million people. The influenza pandemic of 1918-1919 is estimated to have killed 20 million people, about twice the number of soldiers killed in World War I. One important indication of progress is that infectious disease has largely been tamed (although AIDS may prove that statement too optimistic).

But should we now be concerned about a "cancer epidemic"? And is the increase in cancer incidence and deaths related to the large increase in synthetic organic chemicals being produced and used?

Cancer and Chemicals

Is There a Cancer Epidemic?

It is true that the number of people getting cancer has risen, but this is more a measure of success than a cause for concern. Cancer is predominantly a disease of old age; the incidence of cancer is up because so many Americans are living longer. After adjusting for the age of the population, the incidence of cancer has fallen, with the major exception of lung cancer and a few other minor exceptions.[4]

FIGURE 4.1
Total Synthetic Organic Chemicals,
Annual Production

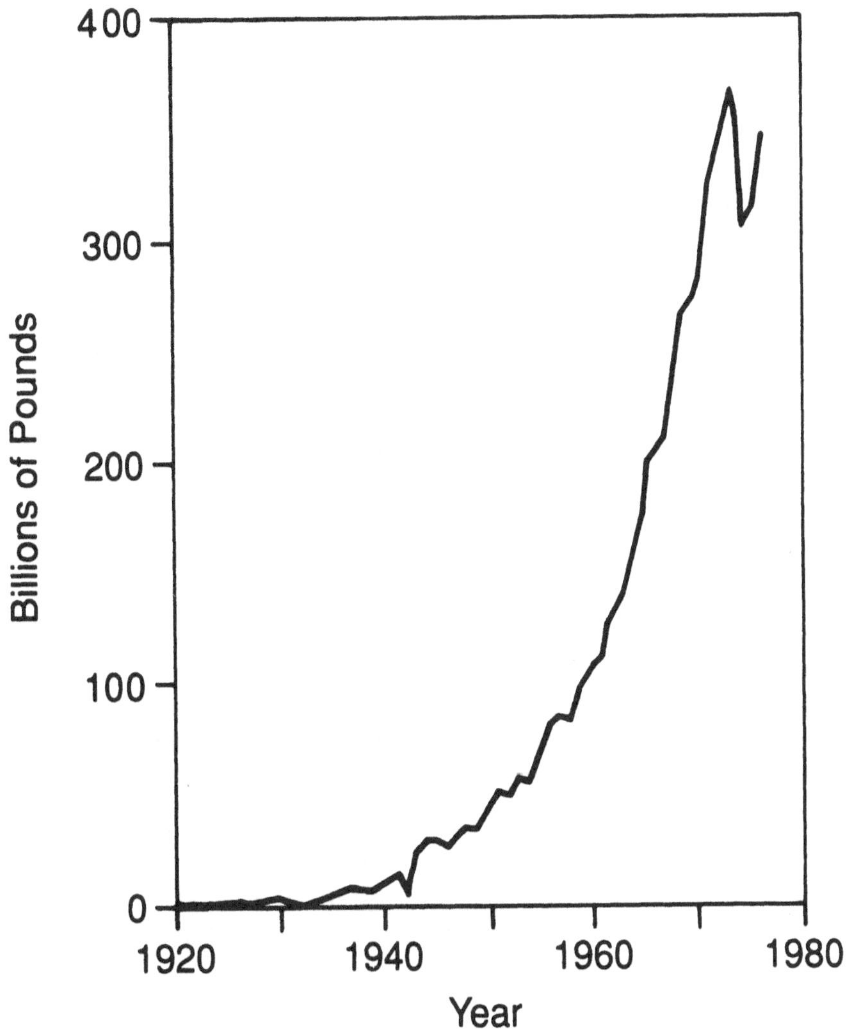

Note: Data exclude tar, tar crudes and primary products from petroleum and natural gas.

Source: Devra L. Davis and B. H. Magee, "Cancer and Industrial Chemical Production," *Science*, 206:1356, 1979. Copyright 1979 by the American Association for the Advancement of Science.

The age-adjusted cancer mortality rate are shown in Figure 4.2. Since 1930, there has been a large increase in lung cancer (presumably due to the increase in cigarette smoking), small increases in leukemia and cancer of the pancreas and prostate, and large decreases in cancer of the stomach, liver, and uterus.

In some cases, increased production of a carcinogenic material has led to increases in cancers. The most important example is the increased use of asbestos, from the shipyards of World War II to the widespread use in buildings and products following the war. The associated rise in mesothelioma and lung cancer demonstrates that high exposure to a carcinogen can result in a "cancer epidemic."[5] If asbestos were the rule rather than the exception, the greater use of chemicals would have led to a spectacular increase in the incidence of cancer.

Pesticides have been a particular concern. Obviously, chemicals designed to kill cockroaches, rats, and fungi are unlikely to produce healthful effects in humans. The concern is accentuated by the fact that the vast majority of pesticides are used in growing and storing food. As expected, pesticides pose acute risks to people from over-exposure; the feared effects of low level, chronic exposure are a different matter. As noted in the 1986 *Annual Review of Public Health*:

> With few exceptions, the delayed effects of pesticides on human health have been difficult to detect. Perhaps the health risks are sufficiently small that they are below the power of epidemiological studies to detect. Yet, it is possible that there are very few effects at all.[6]

Our grandparents' fears of scarlet fever and tuberculosis seem to have been transformed into fears of unperceived poisons from waste dumps or nuclear reactors. Since the level of risk has declined significantly, one might have expected that the level of concern also would decline.

Life Expectancy as a Measure of Risk

Perhaps the best overall measure of risk is the life expectancy of the average American. This measure integrates the various good and bad things happening to people into a single "bottom line." According to this measure, there has been a dramatic decline in risk levels in this century in the United States and throughout the developed nations. Life expectancy continues to

FIGURE 4.2
Age-Adjusted Cancer Death Rates
by Type of Cancer
(United States, 1930-1986)

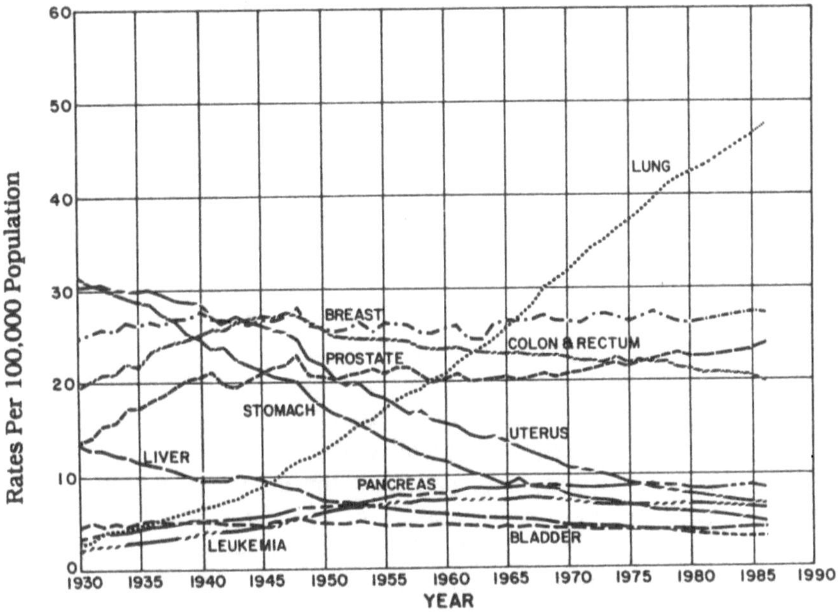

Source: American Cancer Society, *Cancer Facts & Figures—1990* (Atlanta: American Cancer Society, 1990), p. 5.

increase, both life expectancy at birth and at age 40. Thus, although the nature of the hazards and the nature of the harm have changed, there are conclusive data that the overall population risk has declined substantially.

Perhaps the most important reason for the increase in life expectancy has been the increase in income, enabling people to obtain better clothing, shelter, nutrition, education, and health care. Indeed, economic growth has done more to increase health than has medical care. Prosperity "prevents" rickets and other diseases of malnutrition, as well as other killers, such as tuberculosis. Until the sulfa drugs in the 1930s and antibiotics in the 1940s, medicine didn't have powerful drugs to arrest disease and promote health. Yet the infant mortality rate declined, life expectancy rose, and the health of Americans increased between 1830 and 1930.

But prosperity also leads to greater concern. In economic jargon, safety is an "income elastic" or "superior" good; as our incomes rise, we are more attentive to eliminating involuntary risks. Thus, the rise in income is not only the most important force behind improvements in health, it is also the most important source of demands for greater safety.[7]

The Risks and Benefits of Carcinogenic Chemicals

Since cancer is a dread disease, who could argue that carcinogenic chemicals should be produced and used? But the issues are not so simple. Chemicals convey benefits as well as risks.

Some of the drugs used in chemotherapy for treatment of cancer are themselves carcinogens. While physicians would prefer not to use carcinogens, these drugs are the most effective treatment. Taking into account the benefits and potential harm of these drugs, they offer patients the best hope.

Similarly, almost ten percent of women get breast cancer. The most effective method of preventing breast cancer deaths is to use X-rays to detect the disease at an early stage.[8] However, X-rays cause cancer, and so physicians must balance the possible harm of X-rays against the possible benefits of early detection.

Since higher standards of living are perhaps the most important determinant of life expectancy, banning a carcinogenic chemical that has a major influence on economic prosperity could do more harm than good. Lowering income could lead to more disease and earlier death than the possibility of a small number of cancers. Suppose there were a carcinogenic substance with so little potency that it would lead to fewer than one cancer each year if it were used. Suppose further that using the substance would result in a major reduction in food prices or substantially increase the supply of food. Should this carcinogen be banned?

Regulating Risk

Nonetheless, the concern about risks leads to demands that government "do something." For many health problems, little can be done beyond basic medical research. Government can regulate factories that use toxic materials that might expose workers and the public living around the plant to undesirable risk levels. It can also regulate products that discharge toxicants. If people believe that a company or industry is putting them at substantial risk, they can use the power of the government to force that firm or industry to reduce emissions or to shut down.

Each economy exists within a social environment that defines the rules under which a company can operate. Society can give companies virtually complete freedom in conducting their business, as occurred in the decades following the Civil War; or constrain nearly every action, as occurred during World War II. However, using government to regulate the behavior of firms has both good and bad aspects. Popular demands are rarely subtle; they require simple principles that everyone can understand. One such simple principle is banning the use of carcinogens.

The Delaney Clause

In 1958 Representative Delaney proposed that any chemical that causes cancer be banned from food.[9] Congressman Delaney's position was surely on the side of the angels. After all, who could favor adding carcinogenic chemicals to food?

The Food and Drug Administration (FDA) opposed the Delaney proposal initially, arguing that existing legislation gave the agency the power to condemn food that was contaminated or otherwise "unwholesome." The FDA argued that it needed no new legislation to ban carcinogens. In the end, the FDA was persuaded to drop its opposition, since the Delaney Clause was viewed to be redundant, but not restrictive. As it turns out, the Delaney Clause has proven to be far more problematic than anticipated, sometimes requiring the FDA to do something other than keeping the food supply safe and wholesome.

In 1958, there were perhaps a dozen known carcinogens and the ability to detect them was extremely limited. For example, if the manufacturing process for a food color introduced a tiny residual of carcinogenic material, this was undetectable. Analytic chemists could detect perhaps one part per million of a carcinogenic contaminant in the food color.

In the ensuing three decades, the number of known human carcinogens (or groups of chemicals that are carcinogenic) has risen to 26 and the chemicals that cause tumors in rodents, and thus are presumed to cause cancer in people, now number more than 600.[10] The rodent test is sufficiently sensitive (and the selection of chemicals to be tested is sufficiently peculiar) so that two-thirds of the chemicals tested have been concluded to be rodent carcinogens, and thus suspect human carcinogens.

The Development of Risk Analysis

A looming crisis was apparent by the mid-1970s.[11] Scientists at the FDA recognized that they could detect carcinogens at levels that had uncertain health implications for people. For exposures at these trace levels, not only might the number of cancers be small, there might be none at all. FDA scientists began the development of "risk analysis" in order to be able to estimate the potential damage that might be done by these minute exposures.

For most toxic chemicals, the standard presumption is that very low exposures are harmless, even though high exposures can be fatal. For example, vitamin A is an essential nutrient, but very high exposures can cause death. Thus, for most toxicants, scientists search for a "safe" dose. This is done by taking the lowest exposure level that leads to an observed detrimental effect, the LOEL or lowest observed effects level. The LOEL is then divided by a factor of 100 to arrive at a "safe" dose for people.

But the mechanism by which chemicals cause cancer is unknown. For example, ionizing radiation appears to cause cancer by damaging the DNA within cells; thus one "hit" by one particle in one cell is assumed to be the cause of cancer in an individual. The same assumption is generally made for chemicals (particularly "genotoxic" chemicals that damage DNA). A single molecule of the chemical is assumed to interact with a single cell, causing the cell to become malignant and initiating cancer. According to this "one-hit" theory, there is no level of exposure that is safe.

The one-hit theory is accepted for genotoxic agents because it appears plausible and is conservative. But, because the Delaney Clause forbids adding a carcinogen to food, no chemical could be added to food if it was contaminated with even a single molecule of a genotoxic agent. Strictly applied, the Delaney Clause and the one-hit theory would prevent virtually any chemical from being added to food.

Food colors, preservatives, emulsifiers to thicken, and vitamins or dietary supplements generally are contaminated with minute amounts of genotoxic chemicals. But banning these additives would mean that foods would have a high spoilage rate, would be less attractive, and sometimes deadly. Greater risks of food poisoning would be posed by banning sodium nitrite under a strict application of the Delaney Clause.

The Ubiquity of "Natural" Carcinogens

Even these substantial costs might be worth bearing if they resulted in a marked decrease in the incidence of cancer. However, food contains carcinogens that are natural constituents (such as estrogens in veal) and contaminants (such as aflatoxin in peanut butter).[12] During some periods, the levels of carcinogenic contaminants are much higher than the levels of carcinogenic impurities in food additives. For example, summer droughts lead to growth of a fungus that produces aflatoxin in grains and nuts (especially corn and peanuts). Short of condemning the entire harvest, there is no alternative to tolerating some levels of aflatoxin in peanuts, corn products, and milk. But does it make sense to ban a food color with one part per billion of a carcinogenic contaminant while tolerating perhaps one thousand times the cancer risk from aflatoxin contamination?

To survive, plants have had to develop their own pesticides to defend themselves against insects and other pests. Bruce Ames summarizes a great deal of research showing that some of these natural pesticides are carcinogenic.[13] Foods such as mushrooms, sprouts, and veal contain levels of carcinogens much greater than many of the food additives banned by FDA.

FDA developed risk analysis to estimate the risk levels associated with carcinogenic food additives. Since there is a great deal of scientific uncertainty about how much cancer would be produced by various exposure levels, the agency made what it regarded as very conservative assumptions in establishing the risk-analysis process.

The FDA was still faced with the question of what risk level would be considered tolerable. Wanting to be extremely cautious, they chose a safety level of one case of cancer per million people (over their lifetimes). In other words, if the conservative risk-analysis procedure estimated that exposing one million people to the substance would result in less than one cancer, the food additive would be deemed acceptably safe.

Obviously the one-in-one-million safety goal is arbitrary. How conservative it is can be illustrated by Table 4.1. The table shows the incidence of cancer and number of deaths attributable to cancer for one million average Americans. Many people do not realize that almost one person in three can expect to have cancer; about two people in ten will die of cancer.

The FDA states that a food additive would be acceptably safe if, according to their extremely conservative risk analysis, no more than one person in a million exposed to the additive would develop cancer because of exposure. In other words, of one million people consuming the food additive, 300,000 would be expected to get cancer from other causes (e.g., smoking). If the food additive itself would increase the number of expected cancers no more than from 300,000 to 300,001, the FDA would regard this substance as posing an acceptable risk. The last line in Table 4.1 is a reminder that all one million people eventually die of some cause.

The FDA could as easily have chosen a safety goal of one cancer each ten thousand lifetimes or each one billion lifetimes. All of these goals are potentially unobservable; all rely on theoretical models that are regarded as being "plausible" rather than as being "proven." However, the FDA's "one in a million" criterion now appears to be chiseled in stone, even if there is no good rationale for the number.

Cancer Causes

Not only is our ability to predict the likely number of cancers to result from exposure to a carcinogen limited, but we can do little to forecast who might be most susceptible. There is a great deal of variation in the likelihood that an individual will get cancer; genetic inheritance, diet, occupational exposures, and personal habits such as tobacco use have a large influence on incidence of cancer.

The National Cancer Institute asked two distinguished British epidemiologists, Richard Doll and Richard Peto, to estimate the causes of cancer in the United States. They found that the vast majority of cancers are preventable, and that smoking (about 30 percent) and diet (about 35 percent) currently cause almost two-thirds of cancers. Their estimates are presented in Table 4.2.

TABLE 4.1
Cancer Incidence and Mortality Rates
per Million Americans

Disease or Condition	Incidence per 1,000,000
Cancer Incidence	300,000
FDA Safety Goal	1
Cancer Deaths	200,000
Deaths from all Causes	1,000,000

Source: Author's calculations.

Carcinogenic exposure to food additives and industrial products are estimated to contribute less than 1 percent each; pollution is estimated to cause 2 percent, an occupational exposures are estimated to cause 4 percent of cancer deaths. Thus, preventable cancers have little to do with the amounts of chemicals in the environment.

Another bit of evidence supporting this assertion is the ambitious goal of the National Cancer Institute for controlling cancer set out in its 1986 report, *Cancer Control Objectives for the Nation: 1985-2000.*[14] By the end of the century, the Institute's goal is that the number of Americans dying of cancer each year (compared to 1985 levels) be cut in half. This goal is to be attained primarily through prevention, with a focus on smoking, diet, and sexual behavior. Carcinogenic chemicals in the environment are not even mentioned in the report, leading to the conclusion that they contribute so little to the cancer rate that attention should be focused elsewhere.

Thus the FDA goal of not more than one cancer per one million people has to be put in perspective with the estimate that more than 200,000 of the cancers could be prevented by changes in smoking, diet, etc. Furthermore, the estimate of one cancer per million lifetimes is interpreted by many FDA scientists as meaning that no cancers would be expected.[15] The body's immune system is hypothesized to handle small exposures to carcinogens (the possibility of a threshold level). In addition, many

TABLE 4.2
Percentage of Cancer Deaths
Attributed to Various Factors

Cancer Contributors	Percentage of Cancers	Percentage Range
Tobacco	30%	25 to 40%
Alcohol	3	2 to 4
Diet	35	10 to 70
Food Additives	<1	-5[a] to 2
Reproductive and Sexual Behavior[b]	7	1 to 13
Occupation	4	2 to 8
Pollution	2	<1 to 5
Industrial Products	<1	<1 to 2
Medicine and Medical Procedures	1	0.5 to 3
Geophysical Factors[c]	3	2 to 4
Infection	10?	1 to ?
Unknown	?	?

[a] Allowing for a possibly protective effect of antioxidants and other preservatives.
[b] Including timing of pregnancy and menstruation.
[c] Only about 1 percent, not 3 percent, could reasonably be described as "avoidable."

Source: Adapted from R. Doll and R. Peto, *The Causes of Cancer* (Oxford: Oxford University Press, 1981), p. 1256. Reprinted by permission of the Oxford University Press.

of the substances classified as carcinogens are known to cause tumors in rodents, but it is doubtful that all of them cause cancer in humans.[16]

Other federal agencies regulating carcinogens include the Environmental Protection Agency (EPA) (including the offices that regulate exposure to ionizing radiation, carcinogens in the air, carcinogens in the water, and waste dumps), the Occupational Safety and Health Administration (OSHA), the Consumer Product Safety Commission (CPSC), and the Nuclear Regulatory Commission (NRC). EPA recognized the value of risk analysis and took over the major role in developing this tool before 1980.[17] OSHA was reluctant to use risk analysis and continued until 1981 to adhere to the philosophy that carcinogens in the workplace should be banned if feasible; if not, exposure should be reduced to the extent feasible (considering both economic and technical feasibility).

Benzene and "Significant Risk"

The acceptability of risk analysis and what constitutes an acceptable level of safety came to a head in OSHA's regulation of benzene, a chemical known to cause leukemia in humans.[18] In 1978, OSHA issued a regulation requiring a tenfold reduction of worker exposures to benzene. This reduction was deemed to be feasible, both technically and economically (the petroleum industry could pay the costs without fearing bankruptcy).

The benzene regulation led the American Petroleum Institute to sue OSHA. The petroleum trade group argued that the cancer incidence under the old standard was already so small that it was senseless to spend millions of dollars to reduce it further. An expert testifying on behalf of the energy firms said that fewer than one case of leukemia every ten years would be expected among workers exposed to benzene under the old standard. The AFL/CIO sued OSHA claiming that the new standard was not sufficiently protective of workers, since OSHA estimated that some workers would still be expected to get leukemia from the exposure.

After a good deal of legal wrangling, the Supreme Court voided the OSHA standard on the grounds that OSHA had not made a finding that the risk of leukemia under the old standard was "significant."[19] The Court relied on a principle of common law that "the law does not concern itself with trifles." This principle led the Court to insist that an agency show that the risks were above a trivial level before considering regulation. This ruling struck at the heart of OSHA's policy that exposure to a carcinogen was to be reduced to the extent "feasible," interpreted to include both technical and economic feasibility.

By and large, this ruling has been interpreted to apply to all federal agencies; regulators now must make a reasonable showing that a risk is "significant" before they regulate. The same is true for analyses of the risks of injury, such as those conducted by the National Highway Transportation Safety Administration, the Federal Aeronautics Administration, other transport agencies, and the Nuclear Regulatory Commission.

The Delaney Clause and Significant Risk

In 1986, the FDA was sued on the grounds that it was not enforcing the Delaney Clause. Recall that the clause says that no carcinogen can be added to food, while the FDA policy was that carcinogens that conveyed a risk of less than one cancer per million people could be added. The District of Columbia Circuit Court agreed that FDA was not enforcing its statute, even though the court questioned whether the Delaney Clause made sense.[20]

As a result of this decision, FDA must ban a food additive that has any measurable amount (perhaps one part per trillion) of a carcinogen. At the same time, FDA continues to set tolerance levels for natural carcinogenic constituents and for other contaminants in food. Nonetheless, FDA standards are supposed to ensure that the food supply is diverse and inexpensive. The result is that levels of aflatoxin and other natural carcinogens are tolerated that are estimated to cause more than 1,000 cancers per one million people, while food colors are banned that are estimated to cause fewer than one cancer per one million people.

Defining "Significant Risk"

The Supreme Court apparently believed that making the significant risk concept operational would not pose much of a problem; like pornography, it might be hard to define but courts would know it when they saw it. The courts have not had an easy time with the concept of "significant risk," however. In struggling to implement it, a recent ruling by the District of Columbia Circuit Court seemed to introduce a new standard, that of "acceptable risk."[21] However, acceptable risk requires a balancing of risk and benefit, whereas the Supreme Court appears to believe that a level can be defined without regard to benefit. Thus, the courts have further muddied the waters.

Agency Safety Goals

Agencies such as the FDA, EPA and NRC have published explicit safety goals. The FDA continues to rely on a safety goal that views as trivial a risk of less than one in one million. The NRC's risk goal provides that living near a nuclear power plant will not increase the risk of cancer by more than 0.1 percent (one additional cancer for each 1000 current cancers).

In 1988, EPA asked for public comment on four alternative safety goals for a community exposed to a carcinogen: (1) no more than one cancer per million people, (2) one cancer per 10,000 people, (3) one cancer a year in the exposed population, and (4) an *ad hoc* rule that commits the agency to nothing. In a 1989 regulation ordering emissions of benzene to be reduced tenfold, EPA adopted a goal that is a combination of all four proposals.[22] The ultimate goal is no more than one cancer per million lifetimes. In the meantime, no group should be exposed to a risk greater than one cancer per 10,000 people. In applying these two rules, the EPA will also attempt to keep the number of cancers in the population exposed below one per year. Finally, EPA will consider all the circumstances of the case in arriving at a rule, and can violate any or all of the first three principles. For example, in the regulation announcing the new policy, EPA permits the group at greatest risk to be subjected to a risk of one cancer per 5,000 people.

EPA has displayed bureaucratic wisdom in this approach: It commits itself to nothing but sets out some numbers that make several interest groups happy. But this bureaucratic wisdom is bad government, since the statement gives little or no guidance as to what the agency will find acceptable in the future.

Searching for Truth in Past Agency Decisions

Perhaps the best guide to what is a significant risk comes from examining the decisions made by federal regulatory agencies in the past. Figure 4.3 is a graph of 94 decisions where the risk level is the vertical axis and the number of people exposed is the horizontal axis. If there were a simple risk-based rule being used in all these decisions, the graph would show that risks above some level in the graph would be unacceptable.

At first glance, there seems to be little order in the decisions; the agencies might have tossed darts at the graph to arrive at their decisions. However, a few moments of thought shows that the pattern is not random.

Obviously, the individual circumstances of each case matter, and so there is no rigid rule. The greater the risk level (the vertical axis) and the greater the population exposed (the horizontal axis), the greater is the likelihood of agency action.

FIGURE 4.3
Decisions to Regulate by Level of Risk and Population Exposed

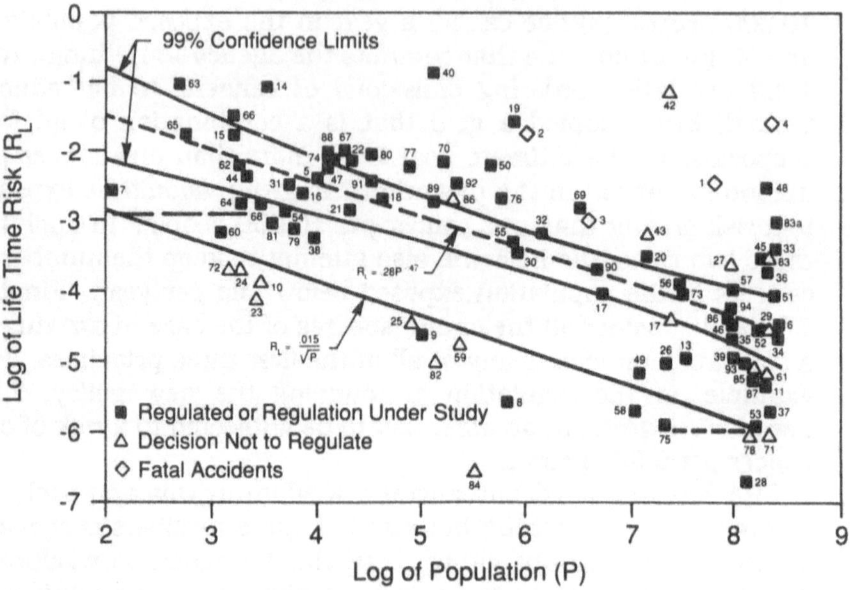

Source: Paul Milvy, "Actual and Perceived Risks from Chemical Carcinogens," *Risk Analysis*, 6:69-80, March 1986.

Key to Figure 4.3[a]

1.	Accident, fatal	— private sector 1982
2.		— mining
3.		— finance, insurance and real estate, 47 years (18-65)
4.		— all, 1982 rate for 70 years
5.	Acrylonitrile	
6.	Alachlor	— dietary
7.		— flaggers
8.		— farmers
9.		— ground applicators
10.	Amitraz	— apple & pear sprayers
11.		— apple & pear consumers
12.		— apple, dietary
13.		— pears, dietary
14.	Arsenic	— copper smelters — high
15.		— copper smelters — low
16.		— glass manufacturing
17.		— Inorganic, neighboring — population average exposure

Key to Figure 4.3 (continued)[a]

18.	Arsenic (cont.)	— maximum exposure
19.	Asbestos	— occupational
20.		— school; students & teachers
21.	Benzene	— fugitive emission
22.		— coke by-product
23.		— maleic anhydride
24.		— ethylbenzene/styrene
25.		— storage
26.		— Stage II gasoline market
27.		— urban
28.		— average population exposure — drinking water
29.		— average population exposure — air
30.	Beryllium	
31.	Butadiene, 1, 3	— occupational
32.	Cadmium	
33.	Captan	— food consumption
34.	Captofol	— food consumption
35.	Carbon tetrachloride	— urban
36.	Chlordane/heptachlor	—food consumption
37.	Chlorobenzilate	— citrus consumers
38.		— citrus applicators (assumed)
39.	Chloroform	— urban
40.	Chromium	— vicinity of large point sources
41.		— urban
42.	Cigarette smokers	— male
43.	Coke ovens	— average exposure for U.S. population at risk
44.		— occupational
45.	Daminozide	— food consumption
46.	1, 2-dichloro-ethane	— urban
47.	Ethylene dibromide	— occupational
48.		— immediate post-regulatory dietary
49.	Ethylene dichloride	
50.	Ethylene oxide	
51.	Folpet	— food consumption
52.	Formaldehyde	— urban ambient
53.		— production use release
54.		— resin manufacturing workers
55.		— apparel workers
56.		— mobile homes
57.		— nonurea/form. homes
58.	Lindane	— shelf paper
59.		— livestock applicators
60.		— pecan applicators
61.		— food
62.	MBOCA	— indirect occupational exposure
63.		— direct occupational exposure
64.		— nonproduction workers (occupational)
65.	MDA	— manufacturing workers
66.		— processing workers

Key to Figure 4.3 (continued)[a]

67.	MDA (cont.)	— all workers — OTS based
68.		— all workers — TLV based
69.	Nickel	
70.	Nitrosamines	— occupational exposure from metal-working fluids
71.	NTA	— Public drinking water
72.		— formulators (occupational)
73.	PCB	— dietary fish
74.	Pentachlorophenol	— applicators/workers
75.		— air
76.	Radiation, ionizing	— all workers in medicine and industry
77.		— power reactor workers
78.		— coal-fired boilers
79.	Radionuclides	— DOE facilities
80.		— uranium mines
81.		— elemental phosphorus plants
82.		— phosphate industry
83.	Radon	— drinking water
83a.		— indoor
84.	Styrene monomer	— occupational
85.	Tetrachloroethylene	— urban air
86.		— dry cleaners
87.	Trichloroethylene	— urban air
88.	Uranium mill tailings	— inactive sites
89.		— active sites
90.	Vinyl chloride	— average exposure for population — Air
91.		— maximum exposure (occupational)
92.		— workers (occupational)
93.		— average exposure for population — water
94.	Volatile Synthetic Organic Compounds	

[a] The values provided in these references reflect current state-of-the-art estimates of risk and population at risk. They are subject to revision and improvement and their inherent limitations should be borne in mind.

For example, workers spraying Amitraz were estimated to have a one-in-10,000 chance of getting cancer; but, since only a few thousand people were at risk, EPA took no action. In contrast, even though the cancer risk from benzene in drinking water is less than one cancer per one million people, EPA took action because essentially all of the U.S. population is exposed.

The figure does not separate risks to workers from those to the general public. However, the agencies are willing to tolerate greater risks to workers, presumably because they are informed about the risks, are paid a compensating differential for the risk, and many fewer people are exposed as workers.

Figure 4.3 and related analyses suggest some general regularities in what agencies view as acceptable risk. First, the greater the population at risk, the smaller must be the risk to any individual. In particular, a hazard such as benzene in drinking water is not acceptable unless it presents a tiny risk to each individual, since more than 200 million people are exposed.

Second, the general public is protected against risk levels that are considered to be acceptable as occupational risk levels. The agencies recognize that workers have some discretion in whether to accept a risky situation; they are better trained than the public to understand it and deal with it and can leave if the situation becomes unacceptable.

Third, the level of safety deemed to be acceptable has been increasing over time. In particular, EPA and OSHA standards for benzene have been tightened.

Fourth, the lack of regularity makes it evident that the individual circumstances in each case are of predominant importance. Thus, a risk goal must be a range, rather than a precise number.

A Proposed Safety Goal

Several researchers have examined how to define risk goals.[23] Dan Byrd and I have proposed to define safety goals using a concept of significant risk: a significant risk is one that is large enough that it could be observed in practice.[24] For example, smoking is a significant risk, as is working on top of a coke oven. Epidemiological studies have shown both practices to lead to large increases in the risk of lung cancer. Occupational exposure to vinyl chloride monomer should be considered a significant risk because three cases of angiosarcoma of the liver, an extremely rare cancer, have been found to be associated with exposure to this substance.

It is difficult to argue that a risk is significant when it is so small that it could not be verified by an epidemiological study. If an extremely rare type of cancer is involved, two cases would be sufficient to verify the risk. Likewise, if a large number of people are exposed, a small increase in risk should be sufficient to generate a detectable risk.

If the agencies were to focus their attention on significant risks, there would be sufficient work to keep the safety and health agencies busy for the foreseeable future. If the agencies perform their work well, we should be able to measure progress in the most meaningful way—a significant reduction in cancer incidence.

Once the agencies have successfully regulated significant risks and reduced them to a nonobservable level, society could then take a hard look at safety and health goals to see where to focus new efforts.

Cost-Effectiveness as a Safety Criterion

In the discussions of safety goals above, significant risk and acceptable risk were mentioned. Significant risk takes no account of the cost of control. For example, if all risks from carcinogens in the environment were reduced to, say, one cancer per one million lifetimes, the costs of achieving the standard would be extremely different across different situations. Achieving a reduction in risk from one cancer per one hundred thousand lifetimes to one cancer per one million lifetimes would cost little per cancer prevented in some situations and be extraordinarily costly in others.

A different way of examining risk goals would be to use a cost-effectiveness criterion. Perhaps the best example of this criterion in agency use is the Nuclear Regulatory Commission's requirement that exposure of workers and the public to ionizing radiation from nuclear reactors must continue to be reduced until the cost of further reduction is greater than $1,000 per man-rem (a unit of exposure to radiation). This criterion translates to a cost of $5-10 million per cancer prevented.

A less controversial way of using cost-effectiveness analysis would be first to estimate the risks of preventing cancer in a wide range of possible government regulations. In some cases, such as radon in buildings, the cost per cancer prevented would be very small. In other cases, such as benzene in drinking water, the cost per cancer prevented would be thousands of times larger.

A cost-effectiveness approach would then instruct the agency to begin with those situations where it was most cost-effective to prevent cancers. The agency would then work its way down the list, using regulation or other means to avert cancer at the lowest cost per case.

The natural question is where an agency should stop. However, in view of the myriad situations facing agency officials, and the difficulty of issuing regulations, this question may not ever have to be faced. There are so many situations to regulate, and it takes so long to regulate, that it is unlikely that an agency would get to regulations that had a cost per cancer prevented that was so high that society would deem it improper.

The Need for Safety Goals

No one likes to be at risk from carcinogens in food, from nuclear reactors, industrial emissions, or the myriad other hazards that surround us. As our income grows larger and other desires are gratified, it is natural to turn more attention to eliminating additional environmental risks. For example, at the end of his life, Howard Hughes is reported to have withdrawn into a special structure that filtered out germs and other common hazards.

In an industrial democracy, the concern for safety gets translated into a demand that the government act to increase safety. While government regulation can increase safety in the regulated situation, it can also slow the growth in income, or even lower income in some cases. Aaron Wildavsky argues that "richer is safer"; misguided attempts at increasing safety via extremely stringent controls wind up lowering income and increasing risk, just the opposite of the goal being pursued.[25]

For most of our decisions, the direct costs of increasing safety, in terms of personal bother, reduced choices, and direct monetary outlays, act as a natural check on losing perspective. However, when a government agency is regulating safety, there appears to be little or no personal cost to the individual who is insisting on greater safety. The direct costs are borne by someone else. Eventually, the regulation will result in higher prices or the loss of some products, but the effects are nonspecific and remote.

Thus, individual concern about risks, and the perception that lowering them generates little or no direct costs, results in public pressure for extremely high safety goals. From a public health perspective, the major cost of this behavior is that resources and attention have been focused on second-order problems, while many of the first-order problems have languished.

Lowering exposure to benzene from five parts per million in air to two parts per billion may have no effect on reducing the incidence of leukemia. This decrease is "estimated" to have an effect because the incidence of leukemia is assumed to be proportional to benzene exposure. But our bodies are able to respond to challenges and to repair damage, at least at low levels. At these extremely low levels, there might well be no health benefit at all from reducing exposure levels.

There clearly are real environmental threats that require attention. In particular, radon in buildings is estimated to lead to perhaps 10,000 to 15,000 cases of lung cancer each year. Radon has received little attention compared to benzene or dioxin, both of which are estimated to cause fewer than 1/100 the amount of cancers.

The economic and social costs of current regulations are high. In its new regulation on benzene, EPA is requiring some companies to spend hundreds of millions of dollars to prevent a single hypothetical cancer. At the same time, EPA is assuring the population that even after this money is spent that cancer risks remain; some people will get leukemia each year from continued exposure to benzene. EPA is drawing such implications from the models it uses, but they are extreme extrapolations based on unproven theory. Predicting cancers at a level of two parts of benzene per billion parts of air is arithmetic extrapolation carried to the extreme.

People want to know how they can lower their risk of having cancer from 30 percent to 10 percent. Regulations such as the benzene rule distract attention from what actions need to be taken and divert resources that ought to be used to achieve major reductions in cancer incidence.

Society desperately needs some social risk goals to help it stop wasting resources on issues that do not need regulation and to help focus attention on issues that do need to be addressed. Restricting regulations to exposures posing a "significant risk"—those that produce an observable effect—has much to commend it. Such a criterion is plausible and would focus resources where they are needed most, eliminating much of the wasted motion resulting from regulation on the basis of unbridled theory.

Notes

1. C. C. Travis, ed., *Carcinogen Risk Assessment* (New York: Plenum Press, 1988).

2. Devra L. Davis and B. H. Magee, "Cancer and Industrial Chemical Production," *Science* 206:1356, 1979.

3. National Research Council, *Toxicity Testing* (Washington, D.C.: National Academy Press, 1984), p. 3.

4. Elizabeth M. Whalen, *Toxic Terror* (Ottawa, Ill.: Jameson Books, 1985).

5. Hans Weill and Janet M. Hughes, "Asbestos as a Public Health Risk: Disease and Policy," in L. Breslow, J. Fielding and L. Lave, eds., *Annual Review of Public Health* (Palo Alto, Cal.: Annual Reviews, 1986), pp. 171-192.

6. D. S. Sharp, B. Eskenazi, R. Harrison, P. Callas and A. H. Smith, "Delayed Health Hazards of Pesticide Exposure, " in Breslow, Fielding and Lave, eds., *Annual Review of Public Health*, pp. 441-471.

7. Aaron Wildavsky, "No Risk is the Highest Risk of All," *American Scientist* 67:32-37, January-February 1979.

8. Lester B. Lave, "Health and Safety Analysis: Information for Better Decisions," *Science* 236:291-295, 1987.

9. Lester B. Lave, *The Strategy of Social Regulation* (Washington, D.C.: Brookings Institution, 1981).

10. Lave, "Health and Safety Analysis," and Lester B. Lave, F. K. Ennever, H. S. Rosenkranz and G. S. Omenn, "Information Value of the Rodent Bioassay," *Nature*, 336:631-633, 1988.

11. P. B. Hutt, "Use of Qualitative Risk Assessment in Regulatory Decisionmaking Under Federal Health and Safety Statutes," in D. G. Hoel, R. A. Merrill and F. P. Perera, eds., *Risk Quantitation and Regulatory Policy* (Cold Spring Harbor, N.Y.: Cold Spring Harbor Laboratory, 1985).

12. Bruce N. Ames, R. Magaw, L. S. Gold, "Ranking Possible Carcinogenic Hazards," *Science* 236:271-280, 1987.

13. Ibid.

14. National Cancer Institute, *Cancer Control Objectives for the Nation: 1985-2000* (Bethesda, Md., 1986).

15. Hutt, "Use of Qualitative Risk Assessment."

16. Lave, Ennever, Rosenkranz and Omenn, "Information Value of the Rodent Bioassay."

17. Elizabeth L. Anderson, "The Risk Analysis Process," in C. C. Travis, ed., *Carcinogen Risk Assessment* (New York: Plenum, 1988), pp. 3-17.

18. J. D. Graham, L. C. Green and M. J. Roberts, *In Search of Safety: Chemicals and Cancer Risk* (Cambridge, Mass.: Harvard University Press, 1988).

19. *Industrial Unions Department, AFL-CIO v. American Petroleum Institute*, 448 U.S.C. 607 (U.S. Supreme Court, 1980).

20. *Public Citizen v. Young*, 831 F. 2d 1108, 113 (D.C. Circuit Court, 1987).

21. *Natural Resources Defense Council v. Environmental Protection Agency*, 824 F. 2d 1146 (D.C. Circuit Court, 1987).

22. *National Emissions Standards for Hazardous Air Pollutants, Benzene Emissions* (Washington D.C.: U.S. Environmental Protection Agency, August 31, 1989).

23. See Daniel Byrd and Lester B. Lave, "Narrowing the Range: A Framework for Risk Regulators," *Issues in Science and Technology* 3:92-97, Summer 1987; Paul Milvy, "Actual and Perceived Risks from Chemical Carcinogens," *Risk Analysis* 6:69-80, March 1986; and Chris Whipple, ed., *De Minimus Risk: Contemporary Issues in Risk Analysis*, Vol. 2 (New York: Plenum Press, 1987).

24. Byrd and Lave, "Narrowing the Range."

25. Wildavsky, "No Risk is the Highest Risk of All."

5

Managing Environmental Risks: What Difference Does Ethics Make?

Margaret Maxey

If indeed the essence of the moral life is choice, the decade of the 1990s promises to be imbued with the right stuff of morality. We need little reminding that a vast sea change has occurred in the way affluent citizens in our high-technology society now interpret its influence on their daily lives. Anthologies have begun to appear denouncing modern technology for having produced "hazard spectaculars." They focus on a litany of technological failures—Love Canal, Three Mile Island, Bhopal, Chernobyl, and the Exxon *Valdez*.

All manner of social complaints are surfacing, leveled against those who are in charge of, and responsible for, a modern "technological leviathan" presumed to be out of control. Critics charge that those who promote new technologies are a new elite who have a vested interest in pushing society toward ever larger and more complex technological systems in which ordinary human values have become eclipsed. When professionals in engineering and the sciences appear to ignore or dismiss such complaints as irrational and emotion laden, it only confirms the impression that those in charge of modern technologies are the stepchildren of a discredited generation of American leadership.

Reduced to simple terms, the focus of complaints seems to be what is derisively referred to as "the old pre-World War II Ethic of Utility"—pursuit of "the greatest good for the greatest number." According to critics, the Ethic of Utility has become not only obsolete but socially unjust. It seems to them that modern technology has been subverted from an instrument for satisfying human needs and has become instead "a license to forget."[1] An autonomous technology, say the reformers, conveys a false message to a younger generation—that human values have been placed on "automatic pilot," as if such a leviathan will run by itself without individual care and moral concern for mitigating social and environmental impacts.

Critics of the old Ethic of Utility insist that it must be replaced by a new Environmental Ethic requiring humans to recapture a mode of living in voluntary simplicity, in closer harmony with a preexisting and fragile balance of Nature. Ethical norms must be derived from harmonies found in an untrammeled environment, sustained by the conviction that "Nature knows best."[2]

The common denominator in this entire array of competing visions is a puzzling anomaly: Why is it that citizens whose standard of living ranks among the highest in the world today, who enjoy political freedoms and social amenities unparalleled in history, have nonetheless developed strongly negative attitudes toward the quality of their lives and the condition of their environment? For centuries human control over a hostile world of Nature has been exerted by those skilled in the principles of engineering and science. These forms of knowledge have been regarded as time-honored sources of safety. But in a remarkably brief period of 30 years, popular confidence has been transformed into skepticism and doubt.

Have innovations in science and engineering during the past half century actually made our world more risk laden, more unsafe? Why do critics protest only those risks imposed by technological hazards, environmental pollutants, and personal contamination caused by unmanageable offspring of an industrialized society? Why are they not concerned about those risks posed in undeveloped and less-industrialized nations? Why should these particular risks be selected for paramount attention? Moreover, what prescribed course of action results from focusing on one set of risks and ignoring others?

Those professionals in government and the private sector responsible for "environmental risk management" will remain vulnerable to charges of squandering public tax revenues and abusing the public trust unless they answer some fundamental ethical questions. First, what cultural processes and criteria explain the current selection of "environmental risks?" Second, what is the merit of scientific evidence used to provide justification for this selection? Third, what ethical considerations are needed to answer the vexing question, "How *fair* is safe enough and clean enough?" as well as to assess the validity of risk management tools such as cost/benefit and cost-effectiveness analysis?

These ethical considerations are by no means exhaustive, but they afford us a sensible point of departure. Let us examine each of these questions in turn, based on the conviction that the immaculate conception of "risks"—conceived as existing "out there" in a physical state of nature, unbegotten by social biases governing their selection—pertains more to the realm of religious belief than to the realm of scientific fact.

Environmental Risk Selection: Moral Criteria

It is tempting to engage in nostalgia about a romanticized period prior to the advent of modern technological advances. Network television programming fuels this nostalgia by dramatizing life with *The Waltons* or *Little House on the Prairie*, not to mention *Grizzly Adams*. Such dramas extol the simplicities of life in the wilderness or in an agrarian society, sustained by friendly neighborhood gatherings in rural communities and by a sense of belonging within small-town community life.

However, no one benefits from historical amnesia, which prevents us from remembering a recent past at the turn of the century when average life expectancy was 40 to 45 years of age. In the early 1900s, more than 13 percent of all infants died before their first birthday. The dreaded disease of the time was not cancer; it was tuberculosis, rivaled by typhoid, diphtheria, smallpox, pneumonia, and a host of other communicable, parasitic diseases causing premature deaths. Farmers protected crops with lead arsenate or "Paris Green" for years before anyone heard of DDT. Red food coloring in common household use was lead chromate, eventually replaced by red dye #3. Red dye #3 was banned by the FDA early in 1990 as a cancer threat.[3]

America has passed through deeply wrenching transitions in its experimental evolution. For a century after the Revolutionary War, our nation remained a rural agrarian republic. After the Civil War, the United States underwent a transition to an industrializing nation. The "Industrial State" followed World War I and the "Technological Society" emerged in the 1960s. In its current expression "the environmental movement" appears to have coincided with this emergence of America's technological preeminence.

Those who chronicle the beginnings of current environmental activism point back to its "first stage" represented by President Theodore Roosevelt and the early Sierra Club. This phase is often described as a reaction to "truly rapacious exploitation of natural resources" in the wake of the Industrial Revolution.

Stage Two was ushered in with the publication of Rachel Carson's *Silent Spring*. In the 1960s people began to be convinced that concern for "the environment at risk" was not to be confined simply to the beauty of Yellowstone Park or Yosemite Valley; rather, all Americans were becoming victims. The view that industrial contamination of water, land and air was sowing seeds of destruction for human beings as well as wildlife became widely held.

In second-stage environmentalism, the strategy adopted has been to enact legislation intended to halt abusive pollution. Some critics of this strategy believe that environmentalism has become relentlessly negative, that it opposes industry by reflex, and prevents legitimate development by driving up economic costs.

However, a different set of critics has identified concerns with "Deep Ecology." They claim that current mainstream environmentalists have become co-opted by the system and are too willing to compromise instead of pushing aggressively for radical change in social values. Whatever their differences, second-stage environmentalists are calling for the displacement of an Ethic of Utility by an Environmental Ethic which derives its norms from an array of assumptions about Nature's fragile balance.

Origin of an Environmental Ethic

The most influential source of environmental assumptions appears to be the 1962 publication of Rachel Carson's *Silent Spring*.[4] In the final chapters of her book, we find four assumptions which appear to have decisively shaped public opinion during the past quarter century.

The first assumption is that Nature has slowly, over eons of unhurried time, reached a delicately balanced evolutionary plateau, fluid and ever shifting, yet nonetheless a "balance of Nature." This equilibrium is held to be increasingly vulnerable to modern man's technological encroachments.

A second assumption is that, during the past quarter century, technological innovations have given modern man a power that can now have planetary impacts both in magnitude and in character. Abuse of this power can have adverse effects on generations yet to come, especially as the result of hazardous chemicals poisoning our groundwater and pollutants destroying the biosphere. By their universal contamination of a natural environment, sinister chemicals are, in partnership with radiation, presumably changing the very nature of life as we know it.

A third presumption is that current rates of cancer are the direct result of exposure to products of post-World War II industries, in particular exposure to synthetic chemicals and radioactive substances. Carcinogens are not natural but man-made; Carson insists, ". . . for man, alone of all forms of life, can *create* cancer-producing substances."[5] Popular belief has echoed the implication that Nature today is fundamentally benign and non-carcinogenic. The Second Report of the Club of Rome in 1974 announced unequivocally at the outset: "The World has cancer, and the cancer is man."[6]

A fourth assumption maintains that repeated small doses of a carcinogen are far more dangerous than a single large dose. A large dose may kill the cells outright, but small doses allow cells to survive, yet in a damaged condition that causes them to mutate and proliferate as alien cells. This is the reason why there is "no safe dose" of any exposure to a carcinogen, in Carson's view. The only safe dose is "zero dose."

Calls for a new Environmental Ethic to displace the old, discredited Ethic of Utility are but the logical culmination of an uncritical acceptance of these four assumptions. The assumptions lead inexorably to the conclusion.

Is there any evidence that Carson's assumptions about the non-natural causation of cancer and of man's role in creating a cancerous universe have been influential in forming social policy? Consider the following examples, provided by Edith Efron.[7]

In congressional testimony while serving as Deputy Administrator of the EPA in 1975, John Quarles clearly echoed Carson's seminal ideas as if they were gospel:

> The great diversity of life in our biosphere reflects the successful resistance of man and other species to the myriad of chemicals found in nature. However, the advent of chemical technology in the past decades has introduced billions of pounds of new chemi-

cals that are often alien to the environment, persistent, and unknown in their interactions with living things.[8]

A second example. In 1976, shortly after the passage of the Toxic Substances Control Act, an anonymous commentator in the *Environmental Law Reporter* demonstrated how entrenched the idea of unnatural chemical carcinogenicity had become:

> Any chemical not found in nature may be said to have the potential to harm biological organisms, from bacteria to man. This is because living organisms' internal defenses and waste removal systems are not likely to be prepared to cope with substances of a kind that they and their evolutionary precursors never had to contend with in the natural universe.[9]

A third example. Umberto Saffiotti of the National Cancer Institute is a major mentor and primary generator of popular assumptions about non-natural carcinogens. In a 1976 paper now regarded as a classic, Saffiotti stated:

> I consider cancer as a social disease, largely caused by external agents which are derived from our technology, conditioned by our societal lifestyle and whose control is dependent on societal actions and policies.[10]

Saffiotti knew at the time that most industrial substances had never been tested for carcinogenicity. Hence, it is significant that he chose the personal pronoun "I" because it signaled a personal belief and not a scientific conclusion derived from a disciplined study of the data. In effect, Saffiotti endowed his personal belief with the status of a "toxicological axiom."

Indeed, Saffiotti went further. He boldly defined the toxicological policy he wished to see applied in the case of inconclusive data:

> The most "prudent" policy is to consider all agents, for which the evidence is not clearly negative under accepted minimum conditions of observation, as if they were positive . . . . In other words, *for a prudent toxicological policy a chemical should be considered guilty until proven innocent.*[11]

Hidden in this policy is the moral imperative it presumes: It must be assumed that no amount of anything labeled a carcinogen is safe, since no one knows the amount of it required to trigger the growth of a malignancy. Even one molecule of any carcinogen must (morally) be presumed to be a potential biohazard.

Here we confront the first paradox besetting the selection and management of environmental risks. With Nature exonerated of harmful health effects, and technological man held culpable, both legally and morally, the magnitude of risks entailed by technological hazards appears overwhelming and yet, paradoxically, reassuring. What man has wrought, man can undo and eliminate. Cancer prevention means purging our natural environment of "billions of pounds" of sinister chemicals and radioactive substances. It is a small wonder that a social selection process has equated environmental risks with man-made toxicants produced by modern technologies and industrial sources.

Summary

Calls for an Environmental Ethic and stringent regulatory controls are the direct result of a social process shaped by deeply entrenched beliefs about moral imperatives and assumptions about Nature. Chief among them is that Nature is noncarcinogenic, virtually benign and benevolent; industrial society and the petrochemical industry are malignant and malicious.

However, the paradox of increasing longevity accompanying technological advances and the assumptions underlying the new Environmental Ethic invite us to explore the scientific evidence. Is the popular belief that "zero pollution" is not only achievable but also a moral imperative based on the facts?

"Zero" Pollution: Evidence and Inconsistencies

The ethical framework governing the selection of environmental risks as moral imperatives for social policy is erected on a scaffolding of beliefs which are seriously at odds with scientific evidence. To be sure, there is an important distinction between "evidence" which persuades and "proof" which convinces. Nevertheless the evidence cannot be denied.

Natural Carcinogens

For reasons which defy a defensible explanation, scientists have only recently turned their professional attention to overwhelming evidence of the pervasiveness of natural carcinogens. Carson told her readers that any natural carcinogens remaining were "few in number" and, anyway, humans had long ago adjusted to their adverse effects. In her book *The Apocalyptics*, Edith Efron has published the most exhaustive inventory of naturally occurring carcinogens to date.[12] She ponders the strange truth that

> within three years of the passage of the Toxic Substances Control Act, conceived to protect the earth from Faustian man, the earth itself had been reported to be carcinogenic beyond anyone's wildest imaginings.[13]

In 1983, a storm of protest greeted Bruce Ames' revealing and provocative article, "Dietary Carcinogens and Anticarcinogens," in *Science*.[14] If any doubt remains, he makes it clear that humans are virtually immersed in a sea of toxic substances inherent in Nature. The human diet contains a plethora of natural mutagens and carcinogens. Consider mushrooms laced with hydrazines, peanuts with aflatoxin, pumpkin pie with myristicin and safrole, and seafoods permeated with toxic microorganisms.

Professor Ames suggests that current preoccupation with relatively small amounts of man-made substances from various industrial processes has diverted scientific research from the most cost-effective ways to reduce cancer, namely by dietary and lifestyle changes. He insists that, if we wish to deal seriously with important causes of cancer, we must of necessity ignore trivia. Professor Ames says, "Thus one can either chase after parts per billion of every man-made carcinogen that turns up or have some sensible regulations about pollution."[15]

Etiology of Cancer

John Higginson's statements some thirty years ago—that most cancer is caused by "environmental factors"—also had a profound effect in establishing the foundations of an environmental ethic. But his statement has been misinterpreted and distorted. Higginson has attempted to correct these distortions by pointing out that, when he ascribed the incidence of cancer in industrialized societies to "environmental causes," he meant the total environment rather than environmental pollutants. These environmental causes are mainly cultural components of lifestyles such as diet and behavior, agricultural practices, hygiene, and social mores.[16]

In a major statement on cancer policy while president of the

National Academy of Sciences, the late Philip Handler stated in a public address:

> . . . we should lay to rest the idea that it is these man-made compounds, abroad in the land, that are responsible for the fact that 25 percent of Americans die of cancer. They are not. The possible effects of all known man-made chemicals, when totalled, could contribute only a miniscule fraction of the total of all carcinogens in our population.[17]

It is useful to reflect on the origin of spontaneous cancer. Studies show that mortality from cancer appears independent of the level of industrialization in a country and, thus, independent of its man-made pollutants. Thus it is not among man-made agents that one should look for primary carcinogens, but instead among all-pervasive "natural" or "normal" environmental components. John Totter suggests that the culprit is oxygen: it is a recognized mutagen; experiments have shown that it causes tumors in fruit flies; in the Ames assay test for screening carcinogens, it shows up positive.[18] Does this mean that Americans should prepare themselves for regulations banning the presence of oxygen above prescribed limits?

Legal Compliance with a Disappearing "Zero"

What a different world we would inhabit if scientists since the 1960s had rejected Carson's assumptions, and instead, had applied to natural carcinogens the same standards of research that were applied to industrial by-products! And what a radically different world we would now inhabit if the Law of Diffusion had been enshrined in the pantheon of regulatory science as an antidote to the quest for zero pollution! This law of physics—which governs the movement and intermingling in nature of all molecules of liquids, gases, and solids—leads to the conclusion that nothing is completely uncontaminated by anything else. Hence the zero-threshold/one-molecule/one-hit theory becomes intellectually bankrupt when confronted with the reality of the Law of Diffusion.

Moreover, it is ethically salient that standards were written in a way that ignored uncertainties in order to simplify legal compliance and enforcement. The result of this pragmatic decision has been a regulatory quest for zero pollution. Thus, the technicalities of legal enforcement have demanded that regulatory standards for pollutants be set at or below the limits of detection.

In the lawyer's world, you are in compliance at any point below the regulatory limit, while you are in violation at any point above that specific limit. Hence, the limit itself is infinitesimal in thickness. As noted by L. B. Rogers, professor of chemistry at the University of Georgia, by stating a regulatory limit in that fashion, the lawyers have repealed the laws of probability—they have failed to recognize that legitimate scientific factors can introduce uncertainties into measurements. This approach makes it easy to write laws, but the result is scientifically untenable.[19]

Moreover, the instruments of detection in today's world of regulatory science have undergone a veritable revolution ignored by lawmakers. There was a time when instrumentation and assay methods could only detect parts per thousand and parts per million. But today's instruments can detect parts per billion and parts per trillion.

Consequently, it is small wonder that analytic scientists have begun to complain about a "receding zero." Lawrence Garfinkle's comments while an epidemiologist at the American Cancer Society reflect the frustration resulting from improved measurement of trace amounts of carcinogens:

> . . . one part per *billion* of anything seems just too small to worry about. Some of us think all instruments capable of detecting chemicals and concentrations lower than one part per million ought to be smashed before we drive ourselves crazy.[20]

Even more trenchant is the statement of former FDA Commissioner Alexander Schmidt. He complained irritably that "we will be chasing a 'receding zero' and some idiot in some lab will come up with something sensitive to parts per quintillion, and our policy says we will adopt it."[21]

Second-stage environmentalists and the general public have yet to fathom the paradox latent in what has haphazardly become "our policy." If the most minute dose, even a single molecule, is assumed capable of causing cancer, then the only "safe" dose or exposure must be zero. And the only way to achieve zero-exposure is to achieve zero-pollution. But sophisticated instruments for detection have demonstrated that "zero" has become an unattainable goal. As a result, compliance, by definition, will always remain an elusive, if not impossible, objective.

It is both the stated and intended goal of environmental activists to pressure regulatory agencies to protect the public from short- and long-term health effects of exposures to mere molecules of toxic substances. If American taxpayers began to understand the ethical inconsistencies and anomalies hidden in an "environmental ethic," would it produce a regulatory backlash?

Consider only three examples of flagrant ethical inconsistencies which make a mockery of fairness in health protection: first, polynuclear aromatic hydrocarbons (PAHs); second, asbestos; and third, radon.

Ethical Inconsistency #1: PAHs

Polynuclear aromatic hydrocarbons (PAHs) have been regulated at 0.03 micrograms per liter in drinking water. Charbroiled sirloin, bread, and lettuce contain detectable levels of a PAH, benzo(a)pyrene. Troyce Jones recently observed that, if the EPA regulated these foods individually at the same intended risk level as it regulates water, then we would only be permitted to eat:

- a 10-ounce steak every two months,
- 2 slices of bread daily, or
- 1.5 pounds of lettuce a day.

"However," he cautions, "each of these estimates would require that one not consume any other food containing PAHs in order to keep the risk from food consumption at the drinking water level."[22]

Ethical Inconsistency #2: Asbestos

It is rarely understood by nonscientists that "asbestos" is a term applied to six naturally occurring minerals used commercially for their desirable physical properties. Chrysotile is the only commercial asbestos mineral belonging to the serpentine group. The other five belong to the amphibole group.

Of the commercial asbestos used in the United States, 96 percent has been chrysotile or "white" asbestos. Evidence demonstrates that this common white asbestos has had the smallest effect on those exposed occupationally. Despite the wide dissemination of white asbestos in our environment—in schools, homes, public buildings, brake-lining emissions, and so forth—there is little evidence that routine non-occupational exposures to this form of asbestos have caused any harm to human beings.

By contrast, a mere 2 percent of the commercially used asbestos is crocidolite, or "blue," asbestos. The documented evidence of its adverse health effects, occupational and non-occupational, is significant. The different health effects of diverse forms of asbestos clearly require different regulatory responses and remedial actions.

Health studies in Canada show that populations can safely breathe air and drink water containing significant amounts of chrysotile fiber. Based on an exhaustive study of epidemiological findings, Malcolm Ross, Research Mineralogist, U.S. Geological Survey, Reston, Virginia, concludes that there is a "threshold" value for chrysotile asbestos exposure below which no measurable health effects will occur.[23] Ross is emphatic about stating dangers of exposure to crocidolite dust and to amosite asbestos. He is equally emphatic about political overreactions and propagandistic uses made of such heavy-handed statements as "One fiber can kill you."

Exaggerated predictions have been made about the amount of asbestos-related mortality expected in the next 20 or 30 years. Political pressures have escalated, calling not only for removal of asbestos from our environment at great cost, but also for reducing or even stopping its use altogether. This is being done even though most asbestos in the United States is of the chrysotile variety. The fact is that asbestos dust levels in schools, public buildings, and city streets are much lower than those found in chrysotile asbestos mining communities where little asbestos-related disease appears in residents generally.

Moreover, the removal of blue asbestos can be dangerous if it is not done with caution and care. In most cases, a lot less environmental disruption—as well as less cost—would result if existing asbestos coatings and insulations were repaired rather than removed.[24] Instead of overreacting to every potential health risk, regulators should allocate both scientific and economic resources equitably—that is, in proportion to the seriousness of environmental health problems.

Ethical Inconsistency #3: Radon

The case of radon in homes compared with asbestos in schools provides still another anomaly in ethical requirements for fairness. On the one hand, EPA has recommended that corrective action be taken in homes where the radon concentration is equivalent to, or greater than, 4 picocuries per liter (4 pCi/L) of radiation. EPA has predicted that there might be 1,300 to 5,000 lung cancer deaths per 100,000 lifetimes among those who live in houses with 4 picocurie concentrations. This hypothetical risk is many times the risk that has been allowed in any occupational setting. Preliminary studies in various parts of the United States indicate that up to 33 percent of the homes have radon levels above 4 pCi/L.

On the other hand, OSHA has made calculations about asbestos exposures based on a working lifetime of 45 years at a level of 2 fibers per cubic centimeter in the ambient air. OSHA estimates that there might be 6,400 cancer deaths per 100,000 lifetimes due to lung cancer, mesothelioma, and gastrointestinal cancer. If this risk is extrapolated to an exposure of 20 years in a schoolroom containing air-borne asbestos at a level of 0.001 fibers/cm^3 the predicted mortality will be 1.4 deaths per 100,000. Similar estimates appearing in a report by the National Academy of Sciences calculate that a 20-year exposure in a school containing 0.001 fibers/cm^3 might be 2.5 deaths per 100,000.

What happens when one compares the risk of radon exposure in homes to asbestos at a level of 0.001 fibers/cm^3 in the schoolroom? Malcolm Ross concludes that it is 400 to 3600 times riskier to live in the 4 picocurie house than it is to work for 20 years in a school that contains 0.001 fibers of asbestos per cubic centimeter in the air.

Having made this comparison, Ross wryly observes:

> If one has children living in a 4 picocurie home, then according to the EPA risk estimates, they would be far safer attending an asbestos-bearing school—provided it does not have radon in the basement. One might consider banning basements, but one would then have to consider banning the first floor.[25]

The point of this analysis is to call attention to the pernicious consequences of a state of affairs in which "environmental risks" are hypothetical, grossly exaggerated, and not compared to other risks. In a misguided quest for zero pollution, so as to achieve zero exposure to mere molecules of man-made pollutants, "environmental risk management" has created a world in which a split personality becomes a mark of sanity.

Perhaps the ultimate irony is the case of nickel. Consider the ethical inconsistency of regulating this naturally occurring carcinogen. Some nickel compounds cause cancer in animals and humans. However, nickel is also considered to be an essential trace nutrient for a number of animal species and may also be essential for humans. If, as expected, the latter case turns out to be true, the federal law that establishes mandatory zero tolerance will in effect also prohibit the presence of a nutrient essential to human life.[26]

These examples illustrate the need to forge an ethical framework different from the current environmental ethic which has proven to be a bottomless quagmire.

Environmental Risk Management: What Difference Does Ethics Make?

A popular misconception about "ethics" can perhaps be traced to busybody preoccupations with the moral behavior of individuals. Reductionist interpretations of ethics have focused both professional and popular attention upon a narrow and virtually exclusive concern about individualistic issues.

However, no one is well served by siphoning public attention away from more aggregated problems having to do with entire organizations and with the validity of patterns of decision making within them. At this level of ethical consideration, a wider lens enables us to focus upon the presence or absence of ethical principles governing consistency and fairness in protecting the best interests of the public at large. Elected legislative bodies, government regulatory agencies, public and private corporations then become the focus of concern.

At a still higher level of concern, issues emerge from considering the merit and social consequences not only of entire world views embodied in socioeconomic systems, but also in value systems which sustain public policies.

Problems encountered in attempts to manage "environmental risks" derive both from conflicting ethical principles at the intermediate, organizational level and competing world views at a societal level.

Thus far I have tried to suggest that, by focusing public attention on "risks" imposed upon a natural environment by industrial/technological man, eloquent critics have reinforced the conviction that "risk" is the only thing that matters. It is now commonplace to believe that today's technology can only threaten our health and safety. Clearly, for whatever reasons, citizens of Western industrialized societies have grown much more risk averse.

Does this phenomenon imply that we are also more "benefits averse" or less sensitive to the connection between deriving benefits from taking risks? Should regulatory institutions be as much concerned about "benefits perception" as about "risk perception" among affected citizens?

The thesis underlying this essay is that decisions governing social policy and establishing regulatory standards derive their ethical justification from the degree to which they are bounded by four ethical plumb lines.

First Ethical Plumb Line: Nature as Ethically Normative

The assumption that Nature is a haven and refuge against the encroachments of modern technology, and that Nature must provide the norms for a new Environmental Ethic leads to another fascinating paradox.

Let us examine more closely the popular assumption that Nature exists in a precarious balance, its fragile condition threatened by unnecessary technologies having an enormous potential to upset that balance, with global consequences.

In the past few years we have grown accustomed to an almost ritual incantation of vivid symbols of technological hazards: Love Canal, Three Mile Island, Bhopal, Chernobyl. Each serves as a reminder that modern technologies have made human life more and more precarious and unsafe. We seem besieged by ominous uncertainties, encroaching on an otherwise safe natural world.

Contrary to this benign view of nature, history is replete with stark lessons about the harm inflicted on humans and their life-sustaining biosphere by the natural environment with its heedless malevolence. The Johnstown flood in 1889 caused 2,209 deaths. Winds from a great hurricane in 1900 caused a storm surge in Galveston, Texas, which claimed 6,000 lives. A volcanic eruption at Mont Pelee, Martinique, in 1902 caused 30,000 immediate deaths. These events surely qualify as "hazard spectaculars." Engineers have devised systems of flood control, as well as early warning systems for hurricanes and volcanic eruptions to attempt to defend against such natural threats.

These tragic reminders of the destructive power of Nature do not exonerate human fallibility nor excuse technological failures, of course. Instead they heighten a sense of our precarious achievements in technological control over Nature's destructive forces. It is specious to argue that, since humans cannot prevent or control the forces of Nature, we ought to concentrate our time, effort and public money on eliminating presumed technological risks. To the contrary, history demonstrates that technological innovations have been remarkably successful in protecting human life from Nature's harmful effects: early warning systems for volcanic eruptions and hurricanes, seismology predictions for earthquakes, dam systems for flood prevention and mitigation all have saved countless lives.

The ethical conclusion is that we can and morally ought to set priorities for health and safety based on cost-effective protection for the most people, whether the source is "natural" or "man-made." Too few analysts recognize the hypocrisy of those who vigorously advocate a new "Environmental Ethic," yet, when proposing policy, refuse to consider the pervasiveness of Nature's toxicants which human beings ingest routinely and without adverse effects. They ignore or reject Paracelsus's dictum: "Dose makes the poison."

Ethics must negotiate the inevitable truth that values conflict, priorities must be set, risks must be taken, and some policies cannot avoid cancelling out others. Before celebrating the birth of a new Environmental Ethic, allegedly derived from values inherent in Nature, and conforming human activities to a presumed "ecosystemic balance," we should be wise enough to recognize that such an ethic "dies the death of a thousand qualifications."

Second Ethical Plumb Line: Environmental Risk as Social Bias

Are regulatory agencies today regulating products that are actually harming people, or are they regulating what frightens people? Should hypothetical risks receive the same priority treatment as actual risks?

In their recent book, *Risk and Culture*, Mary Douglas and Aaron Wildavsky ask, "Of what are Americans fearful?" The answer is both bewildering and ironic: "Nothing much, really, except the food they eat, the water they drink, the air they breathe, the land they live on, and the energy they use."[27]

It is easy to assume, without much reflection, that people differ in their perceptions and selection of risks because their personalities differ. However, Douglas and Wildavsky suggest that the reason people pay attention to certain risks, and ignore others, is purposeful and not arbitrary. That is, people focus on a set of risks in order to reinforce and conform to a preferred way of life and an environmental quality already selected on other grounds. When people disagree about "perceived" versus "actual" risks which they are disposed to accept or avoid, the disagreement signals divergent agendas for changing or preserving preferred forms of social organization. Cultural bias dictates risk selection.

In the current dispute about technological hazards and perceived threats to environmental quality, partisans accuse each other of serving the vested interests of social institutions which are in conflict. One side accuses the other of being a lackey of the "industrial establishment" which is presumed to be the exclusive source of death-dealing pollutants. The other side responds by accusing its critics of joining the ranks of the "danger establishment," dedicated to destroying the very foundations of material well-being that afford them the leisure time to be prophets of doom. Each side accuses the other of irrational bias, of misperceptions of real risks, of subversion of the public interest.

Regulators cannot deal constructively with current disputes over the discrepancy between actual versus perceived risks if they cling to the view that real knowledge of an external world is held by experts in the physical sciences, while mistaken perceptions pertain to the realm of personal psychology. According to this faulty division, dangers are assumed to be inherent in a physical state of affairs; hence the risks posed seem objectively ascertainable only by experts. This division depends upon subjective personality traits (*i.e.*, an individual is either a risk taker or a risk avoider) to explain a bias toward the selection of certain types of risk and not others.

However, an objectivist view is quickly nullified by two questions: Why is it that experts disagree? Why does one and the same individual fear environmental risks to the exclusion of those dangers more immediately life threatening?

Only a cultural theory of risk selection as a product of cultural bias and social criticism can account for the anomalies we encounter in affluent technological societies. The risks we select to control or mitigate, individually and collectively, are a product of the choices we make concerning the best way to organize social relations, to protect shared values, and to devise institutional mechanisms for providing informed consent in the formulation of social policy.

The second ethical plumb line is a clear recognition that risk selection is the result of prior allegiance to a preferred form of social organization and implicit shared values. Cries of "not in my backyard" and demands for "safety" as an antidote to risk taking should not be interpreted as indications of the failure of technology, but instead of its enormous successes in overcoming human subjugation to the destructive forces of uncontrolled nature. It is a matter of historical fact that these successes have generated a level of public expectation that far exceeds the ability of modern technology to achieve, and of our economy to afford. As Aaron Wildavsky sagely remarks:

> . . . the collective urge to risk reduction is so much greater than the sum of individual urges it claims to represent. Each of us would do less for ourselves than we would insist that the government do for us.[28]

Third Ethical Plumb Line: Risk as Rite of Passage to Benefits

One of the most popular misconceptions about "risk" is that it is a concept used to certify consequences which are harmful, dangerous, or "bad." These seem to contrast with consequences which are beneficial, pleasurable, or "good." By implication, benefits become viewed as antithetical to risks or "risk free." This false antithesis between risks and benefits conveys the notion that there is a way to have one without the other.

People typically do not "take risks" because they intend harm to themselves or to others. We hope to benefit from risk taking but this result may not materialize, and instead some harm may occur. Any such harm is clearly an unwanted and unintended side effect. The proper symmetry should be between harms and benefits, not risks and benefits. Risk taking is an unavoidable rite of passage to benefits. Risk is, by definition, conditionally acceptable.

Another popular misconception results from considering a toxic substance, or suspected carcinogen or "hazardous technology" as if it represents only an incremental risk, amounting to a simple addition to current background risk. To the contrary, any "new" risk reorders an entire system by displacing, offsetting, or otherwise restructuring a prior pattern of benefits and harms. Only *systemic* risk analysis enables us to recognize this modification in a meaningful context.

Consider the following examples of risk displacement. Steady economic growth in industrialized, developed societies has resulted in improved housing, better nutrition, environmental sanitation, and more plentiful food from mechanized agriculture. Machines have reduced accidents in the workplace. What potential hazards did people live with before freons made possible refrigeration for food preservation? The incidence of ptomaine poisoning and stomach cancer has dropped dramatically. What potential hazards would we be experiencing if drinking water were not chlorinated? Or what would be the consequences if DDT were not used against malarial mosquitoes or if pesticides, preservatives, and irradiation did not protect our current food supply from insects, mold, botulism, and a host of pathogens?

Unless we make comparisons with preexisting historical conditions, or with naturally occurring risks which would exist in the absence of actions taken by humans, we would totally falsify the fact that "new" risks cause displacements of "old" risks. Congressional legislation has resulted in a regulatory preoccupation with biological health effects predicted from measurements approaching the infinitesimal—without meaningful comparisons to naturally occurring biohazards. This preoccupation cannot do justice to the full spectrum of potential benefits and harms people confront in their daily lives.

From a bioethical perspective, we do ourselves and our posterity a grave injustice by allowing a moral concern for basic rights to health protection to be narrowed down to—and in many cases trivialized by—an obsession with hypothetical health effects from a few toxic substances. This myopic preoccupation siphons public attention away from preventable causes of malnutrition, disease, and premature deaths from natural catastrophes which claim thousands of lives daily in the real world.

Fourth Ethical Plumb Line: Fairness as Equitable Protection

These considerations bring us to the pivotal point of this examination. What is the central ethical question for environmental risk management? Is it, "How safe is safe enough?" Or should it instead be, "How fair is safe enough?"

Steve Rayner and Robin Cantor of the Oak Ridge National Laboratory have recently suggested that concepts of risk have been mistakenly dominated by engineering concerns.[29] As a result, regulators not only reinforce the idea of risks as things "out there" in nature, but also calculate the probability of an adverse event and the magnitude of its consequences in such a way that common concerns of ordinary people have become eclipsed. Rather than probabilities multiplied by consequences, what ordinary people care about are trustworthy institutions, procedures to assure informed consent, and adequate compensation for harmful effects should a project fail.

Different constituencies have different preferences for forms of social organization. These entail vastly different expectations about the proper procedures for securing consent, for judging adequate compensation and liability, and for eliciting trust in institutions responsible for managing potential threats to public health and safety. Rayner and Cantor describe four idealized forms of social organization—first, entrepreneurial individualism functioning in a competitive market system; second, formal organizations functioning in bureaucratic or hierarchical systems of accountability; third, the egalitarian collectivist constituency typified by religious sects or revolutionary political groups; and fourth the constituency comprised of atomized and often alienated individuals. In each general type, Rayner and Cantor uncover the risk conflicts among constituencies which are rooted in disagreements over goals and over principles governing consent, liability, and trust.

Although recognition by regulatory agencies of radical differences among these four types of political constituencies is enlightening, it does little to resolve the difficulty of settling upon any one risk-management solution satisfactory to all who share these diverse cultural biases. Since egalitarians distrust both bureaucrats and entrepreneurial individuals, they favor rights-based justifications for decisions they hope will achieve their goal of a "new social order." Market entrepreneurs favor social-consequences-based justifications, which they hope will achieve their goal of market success. Bureaucrats prefer a structure based on contracts because their goal is maintenance of due process within the existing system.

Inevitably the problem of reaching a compromise consensus in achieving some acceptable measure of public health protection must meet the fundamental criterion of fairness. Ultimately, it is the responsibility of regulators and standard-setting agencies to allocate limited amounts of public revenues equitably, *i.e.*, proportionally, in such a way that they reduce *actual* harm—preventing not only reduced life expectancy but also diminished quality of life.

Regulatory officials cannot escape a necessary comparison of more or less cost-effective methods of reducing threats to the safety and health of taxpaying consumers. Since the percentage of Gross National Product devoted to health care has doubled over the past two decades, citizens are asking the increasingly strident question, "Are we getting our money's worth?" Or are many expensive procedures being undertaken unnecessarily, depriving others of greater benefits if money were spent elsewhere?[30]

The method of cost/benefit analysis was originally developed outside the health care sector and focuses upon economic effects. The method of cost-effectiveness analysis has been developed specifically in relation to health care and focuses upon health effects.

In cost/benefit analysis, all benefits are measured in terms of their monetary worth, using "human capital" calculations which incorporate productivity (wages) gained as a measure of the value of health. By contrast, cost-effectiveness measures health in a variety of ways, for example, decreased length of hospital stays, cases prevented, additional years of life gained, reduction in mortality, and so forth. In managing "environmental risks" that are properly called "public risks," both methods have complementary roles to perform. Demands for "safety" by any constituency which is oblivious to costs is ethically indefensible. It is a cruel distortion to construe these types of analyses as engaging in a measurement of personal expendability in order to achieve technological advancements for an abstraction called "society."

In the real world where priorities must be set and public revenues effectively allocated, regulatory officials are to be commended rather than condemned for comparing the costs of alternative methods of reducing *actual* harm. Merely outlawing hypothetical risks takes little courage and produces only hypothetical benefits. The goal of fairness in a quest for protection of public health and safety should be pursued as a covenantal expression of our common humanity.

Conclusion

As we approach the end of the twentieth century and the second millennium, we must recognize that we live in a technological society which is not about to disappear. With moral righteousness we could denounce it as a Faustian mess already made.

It would be so much easier, of course, to settle back into the mode of living which assumes one's proper role to be that of an objective social critic—relentlessly "picking the eyes out of the potato of life." There is a reassuring sense of satisfaction and solidarity with those eloquent critics who denounce modern technology as an "autonomous force," who lament the exploitive power of a "technocratic elite," and who bemoan "the tyranny of a technological imperative."

The ethical dilemmas posed by highly visible social critics may appear to stem from the nature of modern technology ravaging a pristine, precariously balanced natural environment. But the appearance is more shadow than substance. As a matter of fact, this view derives from our all-too-human propensity to look upon technologies as illegitimate offspring, or in some cases, as convenient scapegoats for a loss of community and fraternal concern for each other.

Like it or not, high technology has bound humans together—nationally and globally—to an extent unprecedented in history. We have never experienced such a strong sense of responsibility for our destiny. This is not a liability but an asset. It brings forth skills worthy of our common humanity.

Notes

1. Langdon Winner, *Autonomous Technology* (Cambridge: MIT Press, 1977), pp. 314-33.

2. Barry Commoner, *The Closing Circle: Nature, Man and Technology* (New York: Bantam, 1972).

3. Steven Waldman, "The Great Cherry Caper," *Newsweek*, February 5, 1990.

4. Rachel Carson, *Silent Spring* (Greenwich, Conn.: Fawcett, 1962).

5. Ibid., p. 195.

6. Alan Gregg, "A Medical Aspect of the Population Problem," *Science* 121:681, 1950, cited as an epigraph in M. Mesarovic and E. Pestel, *Mankind at the Turning Point: The Second Report to the Club of Rome* (London: Hutchinson, 1974).

7. Edith Efron, *The Apocalyptics: Cancer and the Big Lie* (New York: Simon and Schuster, 1984).

8. Ibid., p. 125.

9. "From Microbes to Men: The New Toxic Substances Control Act and Bacterial Mutagenicity/Carcinogenicity Tests," *Environmental Law Reporter*, 6:10251-252, 1976.

10. Umberto Saffiotti, "Risk-Benefit Considerations in Public Policy on Environmental Carcinogenesis," *Proceedings of the Eleventh Canadian Cancer Research Conference* (Toronto: National Cancer Institute, 1976).

11. Ibid. Emphasis added.

12. Efron, *Apocalyptics*, ref. 11, pp. 135-76.

13. Ibid., p. 134.

14. Bruce Ames, "Dietary Carcinogens and Anticarcinogens," *Science* 221:1256-64, 1983; also Bruce Ames, Reply to "Letters," *Science* 224:760 ff, 1984.

15. Bruce Ames, "Water Pollution, Pesticide Residues, and Cancer," adapted from November 11, 1985, testimony to the Senate Committee on Toxics and Public Safety Management. *Water* 27/2:23-24, 1985/1986.

16. John Higginson, "Cancer and Environment: Higginson Speaks Out," reported by T. H. Maugh. *Science* 205:1363-66, 1979/1980.

17. Philip Handler, "Dedication Address," Northwestern University Cancer Center, 1979.

18. John R. Totter, "Spontaneous Cancer and Its Possible Relationship to Oxygen Metabolism," *Proceedings of the National Academy of Sciences*, 1980.

19. L. B. Rogers, cited in "Trace Analysis: Bringing Certainty to Confusion," *The Point Is . . .* (public issues newsletter from Dow Chemical Company) 47:2, 1981.

20. Lawrence Garfinkle, quoted by M. W. Browne, "How Tiny Chemical Traces Are Found," *New York Times*, August 14, 1979, pp. C1-2.

21. Alexander Schmidt, quoted by T. H. Jukes, "DES in Beef Production: Science, Politics and Emotion," in H. H. Hiatt, J. D. Watson, and J. A. Winsten, eds., *Origins of Human Cancer* (Cold Spring Harbor, N.Y.: Cold Spring Harbor Laboratory, 1977), pp. 1658-59.

22. Troyce Jones, "Radiological and Chemical Contamination: Should I Spend Money for the Present or the Future?" *The Health Physics Society Newsletter* XV/5:1-4, 1987.

23. Malcolm Ross, "A Survey of Asbestos-Related Disease in Trades and Mining Occupations and in Factory and Mining Communities as a Means of Predicting Health Risks of Nonoccupational Exposure to Fibrous Minerals," *ASTM Special Technical Publication 834* (Philadelphia: ASTM, 1984) pp. 51-104. See also Malcolm Ross, "A Definition for Asbestos," ibid., pp. 139-47.

24. Ross, "Survey of Asbestos-Related Disease," p. 97.

25. Malcolm Ross, "The Radon Versus Asbestos Risk," unpublished white paper, 1987.

26. Herbert McKee, "Valid Measurements: Key to Decisions," *Chemtech* 25:431-33, 1985.

27. Aaron Wildavsky and Mary Douglas, *Risk and Culture* (Berkeley: University of California Press, 1982).

28. Aaron Wildavsky, "Richer Is Safer," *The Public Interest* 60:23-29, 1980.

29. Steve Rayner and Robin Cantor, "How Fair is Safe Enough? The Cultural Approach to Societal Technology Choice," *Risk Analysis* 7/1:3-9, 1987.

30. Danielle D. Emery and Lawrence J. Schneiderman, "Cost-Effectiveness Analysis in Health Care," *Hastings Center Report* 19:4, pp. 8-13, 1989.

PART THREE

Markets and the Environment

6

Does Capitalism Cause Pollution?

Thomas DiLorenzo

It is widely believed that free enterprise and environmental protection are incompatible. According to this belief, businesses are prone to despoil the environment in their quest for profit. Consequently, government regulation and, in some cases, nationalization of natural resources, are supposedly required to protect the environment.

Such thinking is the basis for recent proposals to greatly expand environmental regulation. So many new controls and regulations have been proposed at the federal level of government that economic journalist Warren Brookes recently forecast that the Environmental Protection Agency (EPA) could become "the most powerful government agency on earth, involved in massive levels of economic, social, scientific, and political spending and interference."[1] Environmental regulations are also proliferating at the state and local levels of government.

If the profit motive and the free enterprise system are the root causes of environmental degradation, however, one would not expect to find much pollution in communist countries, such as the Soviet Union, China, and in Eastern and Central Europe, that for decades outlawed profit making and private enterprise. But exactly the opposite is true: The communist world suffers from the worst pollution on earth. Could it be that free enterprise is not so incompatible with environmental protection after all?

Market Failure and the Environment

Among the first to popularize the idea that private enterprise is primarily, if not solely, responsible for environmental degradation was the British economist A. C. Pigou. In his book *Wealth and Welfare* (1912), Pigou hypothesized that industry is generally unconcerned about the pollution it creates because any damages are regarded as social, not private costs, whereas the manufacturer is only concerned with private costs. "Smoke in large towns which inflicts a heavy loss on the community . . . comes about because there is no way to force private polluters to bear the social cost of their operations."[2] Thus, according to Pigou, a chief cause of environmental spillovers is the divergence between private and social costs. His proposed solution was regulation (or taxation) "in the public interest" that would reduce the spillovers.

Pigou's ideas have been part of mainstream economic thought for decades. Paul Samuelson's textbook, *Economics*, which from 1948 until the mid-1970s was the most popular economics text in the nation (if not the world), taught generations of students that unregulated private enterprise is inherently incompatible with environmental protection. Most other books on environmental economics have conveyed the same lesson.

Not all students of environmental policy stop with mere regulation as a proposed solution. There has long been a corollary belief that environmental misuse would disappear if government were to own the means of production and natural resources. If the economic system was driven by public-spirited bureaucrats instead of profit seekers, environmental harm would, in theory, be minimized because the bureaucrats would take all costs—private and social—into account. If property were government-owned, the interests of the general public supposedly would be protected.

The prominent socialist economist Oscar Lange was among the first to espouse this view. In 1938 he wrote: "A feature which distinguishes a socialist economy from one based on private enterprise is the comprehensiveness of the items entering into the socialist price system."[3] This is also the view of a long line of American environmentalists who for decades have been urging greater regulation and government control and ownership of land and natural resources. As the author of a recent book about the environmental movement has concluded: "The environmental agenda focuses primarily on securing greater government control over natural and environmental resources: more taxpayers' dollars, more regulations, more agencies, and more government ownership."[4]

Some segments of the modern environmental movement devote more of their time and resources advocating regulation rather than outright nationalization of natural resources and the means of production. But there is another segment of the movement that is more extreme. The environmentalist author Barry Commoner exemplifies this extremist strand. "The origin of the environmental crisis," Commoner writes in a recent book, "can be traced back to the capitalist precept that the choice of production technology is to be governed solely by private interest in profit maximization."[5] "Environmental improvement can occur only," says Commoner, if we "implement the social governance of production."[6] He acknowledges that there is a more "doctrinal" term for the "social governance" of production: "socialism, classically defined as social ownership and control of the means of production."[7]

In sum, the theoretical basis for greater governmental regulation, control, and ownership of resources to deal with environmental problems is premised on the belief that private enterprise and profit seeking are the primary causes of environmental externalities and that government ownership or control will "internalize" these social costs. For many, this is the conventional environmental wisdom.

If this theory is correct, one would expect the communist world—where profit making has been outlawed and government has laid claim to virtually all productive resources—to be environmentally pristine. Now that the iron curtain has been lifted, however, the recently revealed facts of severe environmental degradation in the communist world undermine this "conventional wisdom."

Communist Pollution

The Soviet Union

In the Soviet Union there is a vast body of environmental law and regulation that purportedly protects the public interest (see

Box). But the laws and regulations have had no perceivable benefit. The Soviet Union, like all socialist countries, suffers from a massive "tragedy of the commons," to borrow the term used by biologist Garrett Hardin in his classic 1968 article.[8] Where property is communally or governmentally owned and treated as a free resource, resources will inevitably be overused with little regard for future consequences.

The Soviet government's imperatives for economic growth, combined with communal ownership of virtually *all* property and resources, have caused tremendous environmental damage. According to economist Marshall Goldman, who has studied and travelled extensively in the Soviet Union: "The attitude that nature is there to be exploited by man is the very essence of the Soviet production ethic."[9]

A typical example of the environmental damage caused by the Soviet economic system is the exploitation of the Black Sea. To comply with five-year plans for housing and building construction, gravel, sand and trees around the beaches were used for decades as construction materials. Because there is no private property, "no value is attached to the gravel along the seashore. Since, in effect, it is free, the contractors haul it away."[10] This practice caused massive erosion which reduced the Black Sea beaches by 50 percent between 1920 and 1960. Eventually, hotels, hospitals, and, of all things, a military sanitarium collapsed into the sea as the shoreline gave way. Frequent landslides—as many as 300 per year—have been reported.

Water pollution in much of the Soviet Union is catastrophic. Effluent from a chemical plant killed almost all the fish in the Oka River in 1965 and similar fish kills have occurred in the Volga, Ob, Yenesei, Ural and Northern Dvina Rivers. Most Russian factories discharge their waste without cleaning it at all. Mines, oil wells, and ships freely dump waste and ballast into any available body of water, since it is all one big (and tragic) "commons."

Only six out of the twenty main cities in Moldavia had a sewer system by the late 1960s, and only two of those cities made any effort to treat the sewage. Conditions are far more primitive in the countryside.

The Aral and Caspian Seas have been gradually disappearing as large quantities of their water have been diverted for irrigation. And since untreated sewage flows into feeder rivers, they are also heavily polluted.

EXCERPTS FROM THE CONSERVATION LAW OF THE RUSSIAN REPUBLIC (1960)

In the period of the comprehensive building of communism, the economic use of our country's rich natural resources is being intensified and the distribution of production. . . . is being greatly improved. This makes it necessary to establish a system of measures aimed at the protection, rational use and expanded reproduction of natural resources . . . With a view to strengthening conservation . . . the Russian Republic Supreme Soviet resolves that:

Article 1 The following natural resources . . . are subject to state protection: land, minerals, waters, forests, landscapes, resort areas, the animal world, the atmosphere.

Article 2 All lands . . . are subject to protection [including] anti-erosion measures. . . . It is forbidden to use devices . . . that contribute to . . . water and wind erosion of soils.

Article 3 Reserves of solid, liquid and gaseous minerals found in the earth are subject to protection.

Article 4 Surface and underground waters are subject to protection against depletion, pollution and obstruction.

Article 5 Forests are subject to protection . . . . The planning of forestry and lumbering should be based [partly] on the need for forest conservation and reforestation.

Article 6 Wild vegetation is subject to protection and regulated use as a fodder base for domestic and . . . wild animals. . . . The pasturing of livestock should be regulated and carried out without an overuse of pastures.

Article 7 Green plantings in all population centers . . . are subject to protection.

Article 8 Local executive committees are required, in the interests of the present and future generations, to ensure the preservation of . . . virgin nature and picturesque places.

There are 22 Articles in all, mandating additional "protection and regulation" of game sanctuaries, resort areas, animals, and the atmosphere. There are also liability laws: "For unlawful destruction . . . of natural resources . . . persons directly guilty of committing such damage are . . . subject to administrative or criminal liability."

Source: "Current Digest of the Soviet Press," November 30, 1960, cited in M. Goldman, *The Spoils of Progress* (Cambridge: MIT Press, 1972), p. 301.

Some Soviet authorities have expressed fears that by the turn of the century the Aral Sea will be nothing but a salt marsh. One Soviet paper reported that because of the rising salt content of the Aral the remaining fish will rapidly disappear. It was recently revealed that the Aral Sea has shrunk by about a third. Its shoreline "is arid desert and the wind blows dry deposits of salt thousands of miles away. The infant mortality rate [in that region] is four to five times the national average."[11]

The decline in water level in the Caspian Sea also has been catastrophic for its fish population as spawning areas have turned into dry land. The sturgeon population has been so decimated that the Soviets have experimented with producing artificial caviar.

Hundreds of factories and refineries along the Caspian Sea dump untreated waste into the sea and major cities routinely dump raw sewage. It has been estimated that one-half of all the discharged effluent in the USSR is carried in the Volga River, which flows into the Caspian Sea. The concentration of oil in the Volga is so great that steamboats are equipped with signs forbidding passengers to toss cigarettes overboard. As might be expected, fish kills along the Volga are a "common calamity."

Lake Baikal, which is believed to be the oldest freshwater lake in the world, is also one of the largest and deepest. It is five times as deep as Lake Superior and contains twice the volume of water. According to Soviet expert Marshall Goldman, it is also "the best known example of the misuse of water resources in the USSR."[12]

Factories and pulp mills have been dumping hundreds of millions of gallons of effluent into Lake Baikal each year for decades. As a result, animal life in the lake has been cut by more than half over the past half century. Untreated sewage is dumped into virtually all tributaries to the lake.

Islands of alkaline sewage have been observed floating on the lake's surface, including one that was eighteen miles long and three miles wide. These "islands" have polluted the air around the lake as well as the water in it. Thousands of acres of forest surrounding the lake have been denuded, causing such erosion that dust storms have been reported. So much forest land in the Lake Baikal region has been destroyed that some observers report shifting sands that link up with the Gobi Desert; there are fears that the desert may sweep into Siberia and destroy the lake.

In other regions the fact that no compensation has to be paid for land that is flooded by water projects has made it easy for government engineers to submerge large areas of land. As much land has been lost through flooding and salination as has been added through irrigation and drainage in the Soviet Union.[13]

These examples of environmental degradation in the Soviet Union are not meant to be exhaustive but to illustrate the phenomenon of communist pollution. As Goldman has observed, the great pollution problems in the Soviet Union stem from the fact that the government determined that economic growth was to be pursued at any cost. "Government officials in the USSR generally have a greater willingness to sacrifice their environment than government officials in a society with private enterprise where there is a degree of public accountability. There is virtually a political as well as an economic imperative to devour idle resources in the USSR," says Goldman.[14]

China

In China, as in the Soviet Union, putting the government in charge of resource allocation has not had beneficial environmental consequences. Information on the state of China's environment is not as readily available as it is from more open societies, but the information at hand is not encouraging.

According to the Worldwatch Institute, more than 90 percent of the trees in the pine forests in China's Sichuan province have died because of air pollution. In Chungking, the biggest city in southwest China, a 4,500-acre forest has been reduced by half. Air pollution has reportedly caused large crop losses.

There also have been reports of waterworks and landfill projects severely hampering fish migration. Fish breeding was so seriously neglected that fish has largely vanished from the national diet. Depletion of government-owned forests has turned them into deserts and millions of acres of grazing and farm land have been devastated. Over 8 million acres of land in the northern Chinese plains were made alkaline and unproductive during the "Great Leap Forward."

Central and Eastern Europe

With communism's collapse, word has begun to seep out about Eastern Europe's environmental disasters. According to the United Nations Global Environment Monitoring Program, "pollution in that region is among the worst on the Earth's surface."[15] Jeffrey Leonard of the World Wildlife Fund concluded that "pollution was part and parcel of the system that molested the people [of Eastern Europe] in their daily lives."[16] Evidence is mounting of "an environmental nightmare," the legacy of "decades of industrial development with little or no environmental control."[17]

Poland. According to the Polish Academy of Sciences, "a third of the nation's 38 million people live in areas of ecological disaster."[18] In the heavily industrialized Katowice region of Poland, the people suffer 15 percent more circulatory disease, 30 percent more tumors, and 47 percent more respiratory disease than other Poles. Physicians and scientists believe pollution is a major contributor to these health problems.

Acid rain has so corroded railroad tracks that trains are not allowed to exceed twenty-four miles an hour. The air is so polluted in Katowice that there are underground "clinics" in uranium mines where the chronically ill can go to breathe clean air.

Continuous pumping of water from coal mines has caused so much land to subside that over 300,000 apartments were destroyed as buildings collapsed. The mine sludge has been pumped into rivers and streams along with untreated sewage which has made 95 percent of the water unfit for human consumption. More than 65 percent of the nation's water is even unfit for industrial use because it is so toxic that it would destroy heavy metals used by industry. In Cracow, Poland's ancient capital, the gold roof of the 16th century Sigismund Chapel recently had to be replaced because acid rain had dissolved so much of it.[19]

Industrial dust rains down on towns, depositing cadmium, lead, zinc and iron. The dust is so heavy that huge trucks drive through city streets daily spraying water to reduce the dust. By some accounts, eight tons of dust fall on each square mile in and around Cracow each year. The Mayor of Cracow recently stated that the Vistula River—the largest river in Poland—is "nothing but a sewage canal."[20] The river has mercury levels that are three times what researchers say is safe, while lead levels are 25 times higher than deemed safe.

Half of Poland's cities, including Warsaw, do not even treat their wastes, and forty-one animal species have reportedly become extinct in Poland in recent years. While health statistics are spotty—they were not a priority of the communist government—available data are alarming. A recent study of the Katowice region found that 21 percent of the children up to 4 years old are sick almost constantly, while 41 percent of the children under 6 have serious health problems.

Life expectancy for men is lower than it was 20 years ago. In Upper Silesia, which is considered one of the most heavily industrialized regions in the world, circulatory disease levels are 15 percent higher than among the general population, cancer rates are 30 percent higher, respiratory disease is 47 percent higher, and there has been an appalling increase in the number of retarded children according to the Polish Academy of Sciences. Although pollution cannot be blamed for all these health problems, physicians and scientists attach much of the blame to this source.

Czechoslovakia. In a speech given on New Year's Day of 1990, Czechoslovakian President Vaclav Havel said, "We have laid waste to our soil and the rivers and the forests . . . and we have the worst environment in the whole of Europe today."[21] He was not exaggerating, although the competition for the title of "worst environment" is clearly fierce. Sulfur dioxide concentrations in Czechoslovakia are eight times higher than in the United States and half the forests are reportedly dead or dying.[22]

Because of the overuse of fertilizers, farmland in some areas of Czechoslovakia is toxic to more than one foot in depth. In Bohemia, in northwestern Czechoslovakia, hills stand bare because their vegetation has died in air so foul it can be tasted. One report describes the Czech countryside as a place where "barren plateaus stretch for miles, studded with the stumps and skeletons of pine trees. Under the snow lie thousands of acres of poisoned ground, where for centuries thick forests had grown."[23] There is a stretch of over 350 miles where more than 300,000 acres of forest have disappeared and the remaining trees are dying.

A thick brown haze hangs over much of northern Czechoslovakia for about eight months of the year. Sometimes it takes on the sting of tear gas, according to local officials. There are environmental laws, but they are not enforced. Sulfur in the air has been reported at 20 times the permissible level. Soil in some regions is so acidic that aluminum trapped in the clay is released. Scientists discovered that the aluminum has poisoned groundwater, killing vegetation and filtering into the drinking water.

Severe erosion in the decimated forests has caused spring floods in which all of the melted snow cascades down mountainsides in a few weeks, causing further erosion and leading to water shortages in the summer.

In its search for coal, the communist government has used bulldozers on such a massive scale that they have "turned towns, farms and woodlands into coarse brown deserts and gaping hollows." Because open pit mining is cheaper than underground mining, and has been practiced extensively, in some areas of Czechoslovakia the devastation of the land is nearly total.[24]

East Germany. Following the destruction of the Berlin Wall, the East German government reported that nearly 40 percent of the populace was suffering ill effects from pollutants in the air. In Leipzig, half the children are treated each year for illnesses believed to be associated with air pollution.

Eighty percent of East Germany's surface waters are classified as unsuitable for fishing, sports or drinking, and one out of three lakes has been declared biologically dead because of decades of untreated dumping of chemical waste.

Much of the East German landscape has been devastated. Fifteen to twenty percent of its forests are dead and another forty percent are said to be dying. Between 1960 and 1980, at least 70 villages were destroyed and their inhabitants uprooted by the government, which wanted to mine for high-sulfur brown coal. One news account described the countryside as "pitted with moon-like craters" and "laced with the remains of what were once spruce and pine trees, nestled amid clouds of rancid smog."[25] The air in some cities is so polluted that residents use their car headlights during the day and visitors have been known to vomit from breathing the air.

Nearly identical problems exist in Bulgaria, Hungary, Romania, and Yugoslavia. Visiting scientists have concluded that pollution in Central and Eastern Europe "is more dangerous and widespread than anything they have seen in the Western industrial nations."[26]

Public Sector Pollution in the United States

The last refuge of those who would advocate socialistic solutions to environmental pollution is the claim that it is the lack of democratic processes that prevents the communist nations from truly serving the public interest. If this theory is correct, then the public sector of an established democracy such as the United

States should be one of the best examples of environmental responsibility. But U.S. government agencies are among the most cavalier when it comes to environmental stewardship.

There is much evidence to dispute the theory that only private businesses pollute. In the United States, we need look no further than our own government agencies. These public sector institutions, such as the Department of Defense (DOD), are among the worst offenders. DOD now generates more than 400,000 tons of hazardous waste a year—more than is produced by the five largest chemical companies combined. To make matters worse, EPA lacks the enforcement power over the public sector that it possesses over the private sector.

The lax situation uncovered by the General Accounting Office (GAO) at Tinker Air Force Base in Oklahoma is typical of the way in which many federal agencies respond to the EPA's directives. "Although DOD policy calls for the military services to . . . implement EPA's hazardous waste management regulations, we found that Tinker has been selling . . . waste oil, fuels, and solvents rather than recycling," reported the GAO.[27]

One of the world's most poisonous spots lies about 10 miles northeast of Denver in the Army's Rocky Mountain Arsenal. Nerve gas, mustard shells, the anti-crop spray TX and incendiary devices have been dumped into pits over the past forty years. Dealing with only one "basin" of this dump cost $40 million. Six hundred thousand cubic yards of contaminated soil and sludge had to be scraped and entombed in a 16-acre, double-lined wastepile.

There are plenty of other examples of Defense Department facilities that need major cleanup. In fact, total costs of a long-term Pentagon cleanup are hard to estimate. Some officials have conceded that the price tag could eventually exceed $20 billion.

Government-owned power plants are another example of public-sector pollution. These plants are a large source of sulfur dioxide emissions. The federal government's Tennessee Valley Authority operates 59 coal-fired power plants in the Southeast, where it has had major legal confrontations with state governments who want the federal agency to comply with state environmental regulations. The TVA has fought the state governments for years over compliance with their clean air standards. It won a major Supreme Court victory when the Court ruled that, as a federal government enterprise, it could be exempt from environmental regulations with which private sector and local governmental power plants must comply.

Federal agricultural policy also has been a large source of pollution, in the past encouraging overutilization of land subject to erosion. Powerful farm lobbies also have protected "non-point" sources of pollution from the heavy hand of regulation placed on other private industries.

Federal Land Misuse

An important component of the modern environmentalist agenda in the United States is increased governmental land purchases. The government's Land and Water Conservation Fund, the American Heritage Trust, proposals to expand "buffer zones" around national parks, and proposals to prohibit development in and around "wetlands" are directed at nationalizing more and more land.

But government ownership of land as a means of assuring environmental preservation flies in the face of the lessons learned from the tragedy of the commons that has so devastated the communist world. In the United States, as in the Soviet Union, there is a long history of environmental degradation caused by communal property ownership. The much higher degree of political accountability in the United States has limited much of the damage, but has not eliminated it.

The federal government owns more than 60 percent of the lands west of the Mississippi and much of it is managed by the Bureau of Land Management (BLM) of the United States Department of the Interior. One consequence of the government's "management" of these lands has been gross overgrazing by cattle ranchers. Because of their political clout, western cattle ranchers (with the help of their congressional representatives) are able to get the BLM to sell them grazing rights at about one-fourth the cost of the typical private-sector grazing rights. Thus, public grazing lands are overgrazed compared to private properties.

Moreover, since the ranchers do not own the land, they have no economic incentive to make any significant improvements on it. According to the Council on Environmental Quality, the only areas of the United States experiencing desertification are on the Navajo Indian reservation in northeastern Arizona (which is one massive commons) and areas around El Paso, Texas, which are government-owned and have been severely overgrazed. There are hundreds of other examples of the American tragedy of the commons.

Policy Implications

Environmental degradation in the communist world casts doubt on the notion that private enterprise *per se* is the sole or even primary cause of pollution, as so many have believed for so long.

The Transformation of U.S. Liability Law

A basic cause of environmental harm in the communist world—and in the United States for that matter—is a breakdown of the rule of law, not of free enterprise. In all communist states, individual rights are subjugated to the will of the political authorities who claim to have special knowledge of what is in the public interest. Thus, in country after country, communist governments have decreed that their plans for economic growth should supersede individual concerns about pollution. The "common good," they claimed, should not be interfered with by selfish individual interests, such as the desire to breathe clean air.[28]

In the early- and mid-nineteenth century a major change in American liability law with regard to pollution took place. As legal scholar Morton J. Horwitz documents in his book, *The Transformation of American Law, 1780-1860*,[29] American common law upheld the rights of individuals to sue for damages from industrial pollution prior to the mid-nineteenth century. "Natural uses of land were . . . favored . . . by strict liability in tort: any interference with the property of another gave rise to liability," says Horwitz.[30] Class action suits were also common in such instances where entire neighborhoods or towns were subjected to pollution. Thus, individual property rights were protected against what one might call environmental trespassing. Tort law provided strong incentives for the internalization of social costs.

But by mid-century, the courts began to change their views as they increasingly decided that economic growth was in the "public interest" and that individual rights should be subservient to the common good. They began to deny citizens the right to sue for damages of their property due to industrial pollution. State legislatures also outlawed class action suits against industrial polluters.

This transformation of American law had important implications for environmental policy. With the older common law, private enterprises knew they would be held liable for pollution damages. They had incentives to minimize or avoid such damages and, perhaps more importantly, to adopt technology that would limit pollution. With the elimination of their legal liability, however, it was inevitable that fewer resources would be devoted to developing "cleaner" technology.

In short, this transformation of the American common law had a similar effect on the environment as the communist governments' imperatives to pursue economic growth at any cost. In both instances, government authorities decided that individual rights—including the right to be free from pollution damage—should be subservient to the public interest, as defined by the government. It is conceivable that a return to a liability system in which individual polluters are held more responsible for their actions could reduce current pollution problems and provide incentives for industry to develop less polluting technology in the future.

American business has already demonstrated the ability to produce innovative pollution control technology in numerous industries. It is this technology that the Eastern Europeans are hoping to rely on for their own environmental cleanup. American businesses are introducing to Eastern Europe soil-washing machines that separate oil from dirt, more efficient and cleaner power plants, smokestack scrubbers and "baghouses" that take ash out of air emissions, modern landfill technology, and modern wastewater treatment technology, to mention a few examples.

The Golden Goose of Private Enterprise

A second implication of the environmental debacle in the communist world is that a healthy private-sector economy is essential to a country that wishes to clean up its environment. American pollution control technology is of no use to the Eastern Europeans if they cannot afford it. Waste management specialists have estimated that it will cost East Germany alone more than $200 billion just to get its environment up to Western standards.

The problem facing the Soviet Union and the Eastern and Central European countries is that decades of communism has rendered them economic basket cases as well as environmental cesspools. They must first allow private property and free enterprise to flourish in order to produce the wealth required to pay for their environmental cleanup.

This fact has important implications for the United States as well. If, in our zeal to clean up our own environment, we pass laws and regulations that, like so many in the past, provide few environmental benefits but impose enormous costs on the economy, then American industry and workers will suffer competitive setbacks in the global marketplace. Reducing the wealth-creating power of the American economy would then result in *less* funds for investing in a cleaner environment.

Conclusion

The Soviet Union and other communist countries enacted scores of grandiose-sounding environmental laws and regulations decades ago. The laws and regulations were filled with rhetoric about protecting the public interest and the common good, but the results were a "common bad."

The communist experience underscores an inherent problem with environmental regulation. Regardless of how well intentioned regulations may be, they are often ineffective and, at times, even counterproductive. Referring to the lack of incentives for production managers in state-owned firms to adhere to environmental laws, Marshall Goldman writes, "the passage of a law does not mean that it will be enforced, especially when the interests of the governing officials do not coincide with the intent of the law."[31]

It is clear that the environmental extremists are wrong. Capitalism is not the cause of pollution. People pollute, usually because government has weakened or eliminated the normal incentives of private-property rights to limit pollution. Coupled with sound liability laws that hold people responsible for their actions and the enforcement of private-property rights, markets will provide more cost effective "regulation" of the environment when such regulation is indeed needed.

Private enterprise, public accountability, and private-property ownership are essential ingredients of environmental protection. As the historian Paul Johnson recently remarked, "capitalism, being a problem-solving mechanism, could solve the problems of pollution not only technically but commercially, too, by writing environmental protection into its costs. That is, of course, now happening in advanced countries."[32]

Notes

1. Personal interview with Warren Brookes on April 16, 1990.
2. Arthur C. Pigou, *Wealth and Welfare* (London: MacMillan, 1912), p. 159.
3. Oscar Lange and Fred M. Taylor, *On the Economic Theory of Socialism* (Minneapolis: University of Minnesota Press, 1938), p. 103.
4. Jo Kwong Echard, *Protecting the Environment: Old Rhetoric, New Imperatives* (Washington, D.C.: Capital Research Center, 1990), p. 35.
5. Barry Commoner, *Making Peace With the Planet* (New York: Pantheon Books, 1990), p. 219.
6. Ibid., p. 217.
7. Ibid.
8. Garrett Hardin, "The Tragedy of the Commons," *Science*, December 13, 1968, pp. 1244-45.
9. Marshall Goldman, *The Spoils of Progress: Environmental Pollution in the Soviet Union* (Cambridge: MIT Press, 1972), p. 66.
10. Ibid., p. 162.
11. Peter Gumbel, "Soviet Concerns About Pollution Danger Are Allowed to Emerge from the Closet," *Wall Street Journal*, August 23, 1988, p. 22.
12. Goldman, *The Spoils of Progress*, p. 225.
13. Ibid., p. 232.
14. Ibid., p. 188.
15. Cited in Larry Tye, "Pollution A Nightmare Behind Iron Curtain," *The Arizona Republic*, February 25, 1990, p. C-2.
16. Cited in Mike Feinsilber, "Eastern Europe Fighting Worst Pollution in World," *The Chattanooga Times*, January 17, 1990, p. A-10.
17. Tye, "Pollution a Nightmare."
18. Marlise Simons, "Rising Iron Curtain Exposes Haunting Veil of Polluted Air," *New York Times*, April 8, 1990, p. 1.
19. Lloyd Timberlake, "Poland—The Most Polluted Country in the World?" *New Scientist*, October 22, 1981, p. 249.
20. Marlise Simons, "A Green Party Mayor Takes On Industrial Filth of Old Cracow," *New York Times*, March 25, 1990, p. 18.
21. Feinsilber, "Eastern Europe Fighting Worst Pollution."
22. Tye, "Pollution a Nightmare."
23. Marlise Simons, "Pollution's Toll in Eastern Europe: Stumps Where Great Trees Once Grew," *New York Times*, March 19, 1990, p. 11.
24. Marlise Simons, "Central Europe's Grimy Coal Belt: Progress, Yes, But at What Cost?" *New York Times*, April 1, 1990, p. 1.
25. Jeffrey Gedmin, "Polluted East Germany," *Christian Science Monitor*, March 16, 1990.
26. Simons, "Rising Iron Curtain," p. 1.
27. U.S. Comptroller General, *Wastepaper Recycling: Programs of*

Civil Agencies Waned During the 1980s (Washington, D.C.: General Accounting Office, 1989), p. 13.

28. The starkest example of this philosophy is how Stalin literally exterminated "an entire generation of ecologists." See Tye, "Pollution a Nightmare."

29. Morton J. Horwitz, *The Transformation of American Law, 1780-1860* (Cambridge: Harvard University Press, 1977).

30. Ibid., p. 32.

31. Goldman, *The Spoils of Progress*, p. 190.

32. Paul Johnson, "Concern for Environment Has Long History," *Washington Times*, April 20, 1990, p. H-3.

7

Making the Marketplace Work for the Environment

Murray Weidenbaum

There is a high common ground on which economists and environmentalists can meet and thus mutually help to generate more sensible public policy. Several basic ideas, readily supported by both groups can help to attain that common ground.

A strong economy requires a healthy environment. Even the most theoretical economist breathes the same air and drinks the same water as members of the Sierra Club. In fact, he or she may be a dues-paying member.

Also, a strong economy provides the resources for human activity, including dealing with ecological problems. It generates the rising living standard that enables citizens to deal with serious concerns beyond the immediate one of paying for everyday necessities. Balancing economic and ecological concerns is hardly an either/or matter.

Any doubt on that score can be resolved by examining the plight of many Eastern European nations. Their weak economies have been unable to support the environmental cleanup taking place in Western societies. The result has been "an ecological Twilight Zone," where even the snow is black. Fines levied on polluters are ineffective in socialized economies because the government, as owner of almost all property, ends up paying the penalties.[1]

The situation is very different in the United States. Rather than subordinating environmental concerns to economic goals, we tend to subordinate economic considerations to ecological objectives.

Public Support and Individual Reluctance

Every poll of citizen sentiment shows overwhelming support for doing more to clean up the environment. An April 4, 1990 public opinion survey by *The New York Times* and CBS News reported that 74 percent of the sample agreed that the environment should be protected "regardless of cost."[2]

Despite a plethora of new laws and directives by the Environmental Protection Agency (EPA) plus hundreds of billions of dollars of compliance costs expended by private industry, the public remains unhappy with the results. An August 1990 survey, released by Environment Opinion Study, Inc., asked respondents, "In general, do you think there is too much, too little or about the right amount of government regulation and involvement in the area of environmental protection?" Two-thirds answered that there was "too little" involvement, while only one in ten persons believed there was too much.[3]

Unfortunately, most efforts to clean up the environment provide an important example of not wishing to pay the piper. Those same citizens who want environmental improvements "regardless of cost" vociferously and adamantly oppose the location of any hazardous waste facility in their own neighborhood. Nor are they keen on paying for the cleanup. Of course, they strongly favor cleaning up the environment, but each prefers to have the dump site located in someone else's backyard and to have the other fellow pay for it.

An example of this situation is the reaction of the citizens of Minnesota to a $3.7 million grant from the EPA to build and operate a state-of-the-art chemical landfill that could handle hazardous wastes with a high assurance of safety. In each of the sixteen locations that the state proposed, the local residents raised such a fuss and howl that the state government backed off. Ultimately, the unspent grant was returned to EPA.

The Minnesota experience is not exceptional. The EPA was also forced to stop a project to test whether the sludge from a municipal waste treatment plant could be used as a low-cost fertilizer. The public opposition was fierce, even though the EPA was going to use federally owned land and the sludge was expected to increase crop yields by 30 percent.[4]

Since 1980, not a single major new hazardous waste disposal facility has been sited anywhere in the United States. According to a state-by-state review, the outlook for the future is even bleaker, in large part because of a worsening of the emotional atmosphere surrounding any effort to locate a new dump site.[5] As Professor Peter Sandman of Rutgers University has pointed out, the public perceives environmental matters not only emotionally, but also morally. "Our society," he has written, "has reached near consensus that pollution is morally wrong—not just harmful or dangerous . . . but wrong."[6] Yet, the individuals that make up that same public are reluctant personally to assume the burdens associated with that strongly held view.

This ambivalent attitude toward the environment is not new. In 1969, the National Wildlife Federation commissioned a national survey to ascertain how much people were willing to pay for a cleaner environment. At a time of peak enthusiasm for environmental regulation, the public was asked, "To stop the pollution destroying our plant life and wildlife, would you be willing to pay an increase in your monthly electric bill of $1?" The "no" vote won hands down, 62 percent to 28 percent (with 10 percent "not sure").[7] That study, we should recall, was taken before the big run-up in utility bills. Perhaps not too surprisingly, the survey showed strong support for taxing business to finance environmental cleanup.

In other words, most of us Americans very much want a cleaner environment, but we are willing neither to pay for it nor seriously to inconvenience ourselves. We try to take the easy way out—by imposing the burden on "someone else," preferably a large impersonal institution. A brief historical review of how environmental protection has evolved in the United States might be helpful before turning attention to some suggestions for improving current environmental policy.

The Growth of Environmental Protection Activities

Environmental regulation in the United States covers a great variety of concerns—air pollution, water pollution, pesticides, toxic substances, hazardous wastes, drinking water, ocean dumping, noise emissions, and asbestos in schools.[8] The Environmental Protection Agency is the main federal organization operating in this area, and it focuses on issuing rules and approving permits. In recent years, "markets" for pollution control have been created to enable companies to seek less costly approaches to meeting environmental standards. Although the government's efforts continue to focus on cleaning up existing pollutants, industry is increasingly committed to avoiding the creation of pollution, notably by means of recycling.

Early Efforts to Protect the Environment

The notion that environmental protection is a proper function of government did not originate in the twentieth century. Nor did a get-tough attitude toward polluters first begin in the United States. More than 600 years before the National Environmental Policy Act of 1969, the king of England proclaimed a no-nonsense pollution control law, complete with penalties for offenders. Like recent legislative enactments, the king did not consider it necessary to weigh the effectiveness of various deterrents such as fines or emissions fees. Instead, in 1308, he tried the simple and more straightforward strategy of executing the polluters.

Even in the United States, protection of the environment is no newcomer to the realm of government activities. Prior to independence, the Massachusetts Bay Colony enacted regulations to prevent pollution in Boston Harbor. Following the Revolution, most coastal states took some action to ensure that no large floating debris would obstruct navigation of the waterways within their borders.

Throughout the eighteenth and nineteenth centuries and well into the twentieth, local governments bore primary responsibility for the regulation of water and air pollution. Unfortunately, localities found themselves quite helpless to control water pollution coming from upstream, and a shift in the prevailing winds was apt to make a sleepy hamlet the unwilling recipient of smoky particles from a more industrialized town.

By the end of the nineteenth century, the connections between dirty water and contagious diseases had stimulated most states to enact water pollution laws. These early statutes were concerned with the human-health aspects of dirty water rather than with abating pollution. The result was a tendency for the pollution issues to be buried in public health agencies that largely ignored the problem once a disease had been eradicated.

Federal involvement with the environment during the first half of the twentieth century was piecemeal. Antipollution legislation was aimed primarily at keeping interstate and coastal waterways free from debris so as to maintain the flow of navigation. The Refuse Act of 1899 forbade dumping into navigable waters without a permit from the Corps of Engineers. The Oil Pollution Act of 1924 banned oil discharges into coastal waters. Otherwise, protecting health and safety was viewed as a function of the states under their police power.

Modern Regulations Governing Pollution

The first breakthrough in federal water pollution legislation was the Water Pollution Control Act of 1948. The law did little more than provide technical and research assistance to the states, but it demonstrated a national responsibility. The Air Pollution Control Act of 1955 very much resembled the 1948 water pollution legislation. Thus, as recently as 1955, a report of a congressional committee stated:

> . . . it is primarily the responsibility of state and local government to prevent air pollution. The bill does not propose any exercise of police power by the federal government and no provision in it invades the sovereignty of states, counties or cities. There is no attempt to impose standards of purity.[9]

During the 1970s, legislation progressively enlarged the role of the federal government in regulating the environment and committing the nation to ambitious goals. The Environmental Protection Agency was established in 1970 to pull together a variety of scattered activities and provide a unified ecological policy at the national level. EPA now administers programs dealing with air pollution, water pollution, toxic substances, waste disposal, pesticides, and environmental radiation. It possesses an impressive arsenal of powers and duties.

Activities of the EPA center around setting and enforcing standards relating to environmental concerns. The agency defines a standard as "the product of fact and theory provided by scientists, and a public value judgment conditioned by the balance of risks against benefits, with a margin of safety on the side of public health and welfare." EPA has several means of enforcement. Upon finding a violation, it may seek voluntary compliance. If that fails, it can order compliance and take court action. The EPA relies on both criminal and civil penalties.

The Problem: Trying to Do Everything at Once

It is much easier for Congress to express a desire for cleaner air or purer water than for an agency like the EPA to fulfill that desire. To be sure, vast sums of money have been spent for these purposes in recent years. From 1970 to 1990, Congress appropriated over $30 billion for the operation of the EPA (see Figure 7.1). The head count of EPA employment rose from a few hundred in 1970 to over nine thousand in 1990. These numbers are dwarfed by the costs incurred to comply with the government's rules on environmental cleanup—$100 billion a year at present.[10]

These staggering outlays have not prevented the critics from instituting an almost endless array of lawsuits whose main purpose is to get the EPA to act faster and to do more. Typical of the assaults on the EPA is this statement by then congressman and now Governor James J. Florio of New Jersey: "They are not in charge. They do not have the resources by their own actions to get the work done, and they are more interested in cosmetics than anything."[11]

The plaintive response of the EPA administrator at the time was that "EPA's plate is very full right now."[12] The items on that plate are being heaped higher on an almost daily basis. One of EPA's newest responsibilities, for instance, is regulation of genetically engineered pesticides. Moreover, rapid scientific improvements permit the detection and, perhaps, regulation of ever more minute quantities of pollutants.

Meanwhile, John and Jane Q. Public are making the problem worse. In 1960, the average American disposed of 2.7 pounds of waste a day. By 1988, that figure was up to 4.0 pounds each day.[13]

To be sure, the EPA can claim important accomplishments. Between 1970 and 1987, total emissions of sulfur dioxide in the United States fell from 28 million metric tons to 20 million. Carbon dioxide was down from 99 million to 61 million. The decline in lead emissions was more dramatic, from 204 thousand metric tons to 8 thousand. All that occurred despite the rise in population and increased economic activity. Rivers from coast to coast that were nearly devoid of aquatic life teem with fish once again. Lake Erie, so laden with pollutants in 1969 that a river feeding into it caught fire, has been revived.[14]

FIGURE 7.1
Environmental Protection Agency
Spending, 1970-1991

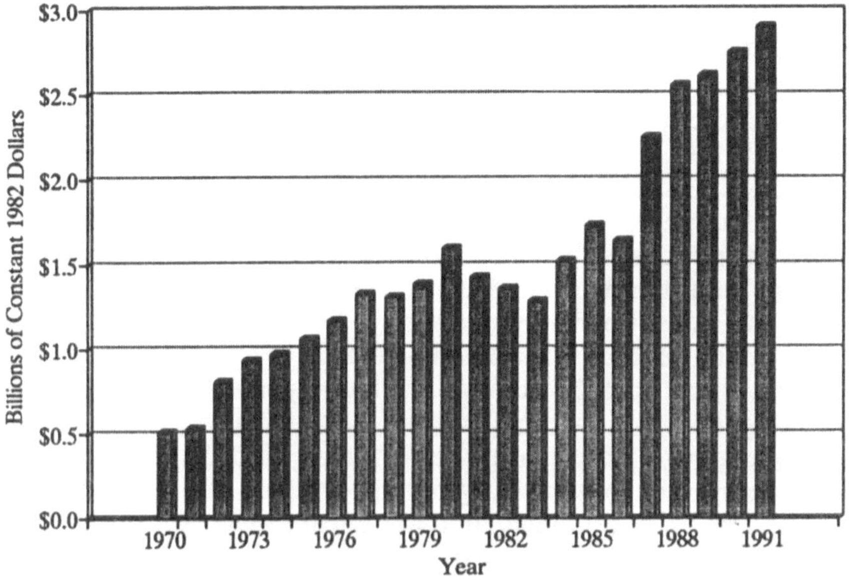

Source: Center for the Study of American Business, Washington University. Derived from the Budget of the United States Government and related documents, various fiscal years.

Despite these successes, the EPA frequently falls short in meeting congressionally mandated goals for pollution cleanup. The hard fact is that the status quo in environmental policy is not sufficient to ensure steady progress at a price the economy can afford. Congress continues to pass high-sounding legislation with unrealistic timetables and inflexible deadlines, while the EPA gets ever greater responsibility and private industry spends billions more on environmental compliance. In the words of the EPA's former administrator William Ruckelshaus, "EPA's statutory framework is less a coherent attack on a complex and integrated societal problem than it is a series of petrified postures."[15]

The Failure to Tackle First Things First

A 1987 report by EPA's Science Advisory Board concluded that the agency's priorities "do not correspond well" with its rankings by risk of the various ecological problems that it is managing, and little has changed since. A 1990 report by the Board's Relative Risk Reduction Strategies Committee concludes that ". . . at EPA there has been little correlation between the relative resources dedicated to different environmental problems and the relative risks posed by those problems."[16]

The EPA's 1987 study found areas of high risk but little regulatory effort. A key example is runoff of polluted water from farms and city streets. Conversely, the study showed that areas of "high EPA effort but relatively low risks" included management of hazardous wastes, cleanup of chemical waste dumps, regulation of underground storage tanks containing petroleum or other hazardous substances, and municipal solid waste. The reason for this mismatch between needs and resources is obvious. The EPA's priorities are set by Congress and reflect public pressure more than scientific knowledge. Driven by the forces of environmental politics, the nation has repeatedly committed itself to goals and programs that are unrealistic. This has meant deploying regulatory manpower unwisely and diverting limited resources to concerns of marginal importance.

The results of this mismatch are substantial. Not all hazards are created equal. Some disposal sites are being filled with innocuous material while truly dangerous substances are or will be, for lack of space, dumped illegally or stored "temporarily." What would help is more widespread application of the legal concept known as *de minimis non curant lex*—the law does not concern itself with trifles.

Back in 1979, a federal circuit court supported the view that there is a *de minimis* level of risk too small to affect human health adversely. It cited that doctrine in turning down the claim that a minute "migration" of potentially carcinogenic substances from packaging into a food product constituted a violation of the Delaney Clause of the Food, Drug and Cosmetics Act. In 1985, the FDA concluded that using methylene chloride to extract caffeine from coffee presented a *de minimis* risk. Hence, the substance is safe for its intended use. In 1987, the National Research Council recommended that the EPA apply a "negligible risk" standard across the board in determining how much of which pesticides can be permitted to show up as residues in fresh produce.[17]

The Public Sector Drags Its Feet

Another part of the pollution problem is that misconceptions of the villains abound. Many people fall into a common trap—that of associating polluters exclusively with business. Many companies do generate lots of pollution. But the same can be said about government agencies, hospitals, schools, and colleges.

Moreover, the EPA lacks the enforcement power over the public sector that it possesses over the private sector. Reports of plants closing because of the high cost of meeting environmental standards are common. In contrast, there is no record of a single government facility closing down because it was not meeting ecological requirements.

It is not surprising, for instance, that the General Accounting Office (GAO) says the performance of federal agencies in carrying out the requirements of hazardous waste disposal "has not been exemplary." A GAO report issued in 1986 says that, of 72 federal facilities inspected, 33 were in violation of EPA requirements and 22 had been cited for Class 1 (serious) violations. Sixteen of the 33 facilities remained out of compliance for six months or more. Three had been out of compliance for more than three years.[18] A follow-up report by the GAO in 1987 showed little further progress. Only four of eleven federal agencies had completed the identification of hazardous waste sites and none had finished assessing the environmental problems they had uncovered. Of 511 federal sites failing to meet EPA standards, only 78 had been cleaned up.[19]

A major offender is the Department of Defense, which now generates over 500,000 tons of hazardous waste a year. That is more than is produced by the five largest chemical companies combined.[20] The lax situation uncovered by the GAO at Tinker Air Force Base, in Oklahoma, is typical of the way in which many federal agencies respond to the EPA's directives. The GAO report stated "Although DOD [Department of Defense] policy calls for the military services to . . . implement EPA's hazardous waste management regulations, we found that Tinker has been selling . . . waste oil, fuels, and solvents rather than . . . recycling. . . ."[21]

The General Accounting Office also reported that two of the five commercial waste sites receiving the base's wastes had major compliance problems. Also, personnel at Tinker Air Force Base were dumping hazardous wastes in landfills that themselves were in violation of EPA requirements. In one case, the

EPA had been urging the Oklahoma Department of Health for several years not to renew the landfill's permit. In another instance, the State Water Resources Board was seeking a court order to close the site. Public agencies, including those in state and local governments, continue to be reluctant to follow the same environmental standards that they impose on the private sector.[22]

Economic Solutions to Environmental Problems

Dealing with Hazardous Wastes

Turning to specific environmental problems, we can start with the controversy over the disposal of hazardous wastes. Instances of toxic waste contamination at Love Canal, in New York State, and at Times Beach, Missouri, have brought a sense of urgency to the problem. The public mood on the subject of hazardous waste leaves little room for patience—but much opportunity for emotional response.

Emotionally charged responses are encouraged by the fact that even scientists know little about the effects on human health of many toxic substances, such as the various forms of dioxin. Levels of some substances can now be measured by the EPA in terms of parts per billion and occasionally per quadrillion, but even the experts still debate the significance of exposure at those dosages. In effect, the scare headlines about chemical health hazards deal with exposures that are akin to the proverbial needle in the haystack. Actually, the needle-haystack comparison is much too modest. One part per billion is the equivalent of one inch in 16,000 miles, a penny in $10 million, four drops of water in an Olympic-size pool, or a second in thirty-two years.

The most severe reaction to dioxin reported so far by humans is a bad case of chloracne, a severe acne-like rash. The bulk of the available information on dioxin and other hazards is based on extrapolating from data on animal experiments, which is very tricky. Most tests on animals are conducted at extremely high concentrations of the suspected element, which do not reflect real-world conditions in which the animals (or humans) live. Scientists note that the massive doses that are fed the animals overwhelm their entire bodies. Moreover, a level of exposure that is harmful to one type of animal may not be injurious to another. For example, the lethal dose for hamsters of the most toxic dioxin (2, 3, 7, 8 TCDD) is 5,000 times higher than that for guinea pigs.[23] Extrapolating the results to humans is even more conjectural.

However, the people in Times Beach, Missouri, and in Love Canal, New York, did suffer severe financial and psychological damage, not from hazardous wastes but from the scare stories they saw and heard so frequently.

In trying to avoid a repetition of these situations, the EPA has promulgated detailed regulations on how polluters must keep track of hazardous wastes and how they should dispose of them. Yet, despite all this effort and attention, the problem of how to dump hazardous wastes is scarcely less serious than it was in 1980, before Congress passed the original Superfund law.

As it stands, the law provides for a large fund raised primarily through taxes on producers of chemical and petroleum products. The EPA uses this money to identify and clean up hazardous waste sites. But little progress is made because, as noted earlier, there is a severe shortage of dump sites.

A more clearheaded view of waste disposal problems is needed in the United States. Because definitions vary among levels of government, estimates of the amount of hazardous waste disposed of each year in the United States range from 30 million to 264 million metric tons. Most of this waste is buried in landfills because incineration, the safest and most effective means of disposal, is nearly ten times as costly. Even so, government and industry spend over $5 billion each year to manage toxic wastes.[24]

Many experts believe that using landfills for hazardous wastes is inherently unsafe, if for no other reason than that they are only storage sites. Moreover, there are not enough of them. The EPA estimates that 22,000 waste sites now exist in the United States, and fully 10 percent of them are believed to be dangerous and leaking.

The result: not enough reliable, environmentally safe places to dump toxic substances. Although EPA wants to clean up as many landfills as possible, it has very little choice as to where to put the material it removes under the Superfund mandate. Taxpayers may wind up paying for the costly removal of waste from one site, only to find later on that they have to pay again for removing it from yet another dangerous site.

Meanwhile, legal fees mushroom. The litigation costs involving cleanup at the various Superfund sites are estimated to run somewhere between $3.5 billion and $6.4 billion—perhaps one half of the total funds devoted to cleaning up.[25]

Overcoming the NIMBY Syndrome

Eventually, society will have to face the main reason for the scarcity of hazardous waste sites—the "Not In My Backyard" (NIMBY) syndrome. Sites for the disposal of toxic substances have joined prisons and mental hospitals as things the public wants, but not too close by.

The hazardous waste disposal problem is not going to disappear unless Americans change to less-polluting methods of production and consumption. Until then, greater understanding is needed on the part of the public and a willingness to come to grips with the difficult problems arising from the production and use of hazardous substances. Of course, it will cost large amounts of money to meet society's environmental expectations. But spending money may be the easiest part of the problem. Getting people to accept dump sites in their neighborhoods is much more difficult.

The answer surely is an appeal not merely to good citizenship but also to common sense and self-interest. In a totalitarian society, people who do not want to do something the government desires are simply forced to do so, with the threat of physical violence ever present. In a free society with a market economy, we offer to pay people to do something they otherwise would not do. The clearest example in modern times is the successful elimination of the military draft coupled with very substantial increases in pay and fringe benefits for voluntarily serving in the armed forces.

Individual citizens have much to gain by opposing hazardous waste facilities to be located near them—and there is a basic logic to their position. It is not fair for society as a whole to benefit from a new disposal site, while arbitrarily imposing most of the costs (ranging from potential health threats to depressed property values) on the people in the locality. But local resistance to dealing with hazardous wastes also imposes large costs on society as a whole. Those costs are in the form both of inhibiting economic progress and having to ship waste from one temporary site to another.

Economic Incentives. There is a way of reconciling individual interests and community concerns: by providing economic incentives. The idea is to look upon environmental pollution not as a sinful act but as an activity costly to society and amenable to reduction by means of proper incentives. After all, the prospect of jobs and income encourages many communities to offer tax holidays and other enticements to companies considering the location of a new factory—even though it may not exactly enhance the physical environment of the region.

Under present arrangements, however, there is little incentive for the citizens of an area to accept a site for hazardous wastes in their vicinity, no matter how safe it is. Unlike an industrial factory, a hazardous-waste facility provides few offsetting benefits to the local residents in the form of jobs or tax revenues.[26] But perhaps some areas would accept such a facility if the state government (financed by all the citizens benefiting from the disposal facility) would pay for something the people in that locality want but cannot afford—such as a new school building, firehouse, or library or simply lower property taxes.[27]

Classify Wastes Correctly. There is much that government can do to improve environmental policy in other ways. The EPA could reduce the entire hazardous waste problem by distinguishing between truly lethal wastes—which clearly should be disposed of with great care—and wastes that contain only minute amounts of undesirable materials. To the extent that changes in legislation would be required, the agency should urge Congress to make them.

The experience of a company in Oregon provides insights into why Congress needs to legislate common sense into the antipollution laws. The firm has been dumping heavy-metal sludges on its property for over twenty years. Company officials told the General Accounting Office that they automatically classify the material as hazardous. Why? Because it would be too costly and time-consuming to try to prove that it was not. The GAO learned from several industry associations that other companies, similarly uncertain and wanting to avoid expensive testing costs, simply declare their wastes to be hazardous, whether they really are dangerous or not.[28] That is not the only example in which those complying with environmental regulations lose sight of the fundamental objectives to be met.

Cancerphobia Misallocates Resources

Another approach to eliminating the gridlock in regulatory policy is to focus on the underlying public concern that is driving the pressures for more sweeping environmental and other social regulation. That concern is the worry about cancer. The regulatory waters have become badly muddied by the public's misconception of the causes of cancer. A widely held notion is that the environment is primarily responsible. There is, of course, a germ of truth to that belief.

It turns out that several years ago a distinguished scientist—John Higginson, director of the World Health Organization's International Agency for Research on Cancer—assigned the primary blame for cancer to what he labeled "environmental" causes. His highly publicized finding that two-thirds of all cancer was caused by environmental factors provided ammunition for every ecological group to push for tougher restrictions on all sorts of environmental pollution.

However, upon a more careful reading, it is clear that the eminent scientist was referring not to the physical environment but to the age-old debate of "environment" versus "heredity" as the main influence on human beings. In the case of cancer, he was identifying voluntary behavior—such as personal life-styles and the kinds of food people eat—as the main culprit responsible for cancer. Dr. Higginson later pointed out, "But when I used the term environment in those days, I was considering the total environment, cultural as well as chemical . . . air you breathe, the culture you live in, the agricultural habits of your community, the social-cultural habits, the social pressures, the physical chemicals with which you come in contact, the diet, and so on."[29] But that explanation has not slowed down the highly vocal ecology groups who latched on to a "catchy," albeit confused, theme—environmental pollution causes cancer.

More recently, one university scientist tried to add some objectivity to the cancer debate by quantifying the issue. Professor Harry Demopoulos of the New York Medical Center examined why approximately 1,000 people die of cancer each day in the United States. About 450 of the deaths, or 45 percent, are attributable to diet. Citing the work of Dr. Arthur Upton of the National Cancer Institute, Demopoulos noted that eating more fresh fruits and vegetables and curtailing fat consumption would be most helpful.[30]

The second major cause of cancer deaths, according to Demopoulos, is the consumption of excessive quantities of distilled liquor and the smoking of high-tar cigarettes. These voluntary actions resulted in 350, or 35 percent, of the cancer deaths. But this is not the environmental pollution that motivates most ecology activists.

A distant third in the tabulation of leading causes of cancer is occupational hazards, accounting for 5 percent of the total. Demopoulos believes that this category may have leveled off and be on the way down. He reasons that many of the occupationally induced cancers are due to exposures two or more decades ago, when scientists did not know that many chemicals were capable of causing cancer.

A fourth category, accounting for 3 percent, is caused by exposure to normal background radiation. The fifth and last category of causes of cancer (accounting for 2 percent) is preexisting medical disorders, including chronic ulcerative colitis, chronic gastritis, and the like.

The remaining 10 percent of the cancer deaths in the United States are due to all other causes. It is noteworthy that air and water pollution and all the other toxic hazards that are the primary cause of public worry are in this miscellaneous 10 percent, not in the 90 percent. Government policy is unbalanced when the great bulk of the effort deals with a category of risk that is only some fraction of one-tenth of the problem.

Hard data can dissipate much of the fear and fog generated by the many cancer-scare stories that the public has been subjected to in recent years. Overall, cancer death rates are staying steady or coming down. The major exception is smoking-related cancer. For the decade 1974-83, stomach cancer was down 20 percent, cancer of the cervix-uterus was down 30 percent, and cancer of the ovary was down 8 percent.

Life expectancy is steadily increasing in the United States (to an all-time high of seventy-five, for those born in 1985)[31] and in most other industrialized nations, except the Soviet Union. This has led the cancer expert Professor Bruce Ames of the University of California to conclude, "We are the healthiest we have been in human history."[32] That is no justification for resting on laurels. Rather, Ames's point should merely help lower the decibel level of debates on environmental issues and enable analysis to dominate emotion in setting public policy in this vital area.

A Birth Control Approach to Pollution

Over 99 percent of environmental spending by government is devoted to controlling pollution after it is generated. Less than 1 percent is spent to reduce the generation of pollutants.[33] For fiscal 1988, the EPA budgeted only $398,000—or 0.03 percent of its funds—for "waste minimization." That is an umbrella term that includes recycling and waste reduction.[34]

The most desirable approach is to reduce the generation of pollutants in the first place. Economists have an approach that is useful—providing incentives to manufacturers to change their production processes to reduce the amount of wastes created or to recycle them in a safe and productive manner.

The Hazardous Waste Example Revisited

As noted earlier, the government taxes producers rather than polluters. By doing that, the country misses a real opportunity to curb actual dumping of dangerous waste. The federal Superfund law is financed with taxes levied on producers of chemical "feedstocks" and petroleum plus a surtax on the profits of large manufacturing companies and contributions from the federal Treasury. Thousands of companies outside of the oil and chemical industries wind up paying very little, whether they are large polluters or not. Contrary to widely held views, a great deal of pollution occurs in sectors of the economy other than oil and chemicals. The manufacture of a single TV set generates about one hundred pounds of toxic wastes.[35]

Switching to a waste-end fee levied on the amount of hazardous wastes that a company actually generates and disposes of would be far more economically sound than current practices. This more enlightened approach would require a basic correction in the Comprehensive Environmental Response, Compensation, and Liability Act (or "Superfund"), but it would be a very beneficial form of hazardous waste "birth control."

A General Application of Market Incentives

More generally, if the government were to levy a fee on the amount of pollutants discharged, that would provide an incentive to reduce the actual generation of wastes. Some companies would find it cheaper to change their production processes than to pay the tax. Recycling and reuse systems would be encouraged. Moreover, such a tax or fee would cover imports which are now disposed of in our country tax-free. In short, rewriting statutes, such as the Superfund law, so that they are more fair would also help protect the environment—and would probably save money at the same time.

Already, some companies are recycling as they become aware of the economic benefits.[36] For example, one chemical firm burns 165,000 tons of coal a year at one of its textile fibers factories, generating 35,000 tons of waste in the form of fly ash. The firm recently found a local cement block company that was testing fly ash as a replacement for limestone in making lightweight cement blocks. The chemical company now sells the fly ash to the cement block manufacturer. What used to be an undesirable waste by-product has been turned into a commercially useful material. Simultaneously, the companies are conserving the supply of limestone.

Creative solutions can at times convert a cost into a benefit. For example, an Allied-Signal Corporation plant in Metropolis, Illinois, created a veritable sea of calcium fluoride sludge as a by-product of its manufacture of fluorine-based chemicals. It was generating the sludge at the rate of 1,000 cubic yards a month. Analysis showed that the sludge could be mixed with another waste stream to produce a reaction. The result: neutralization of the waste and production of synthetic fluorspar that Allied uses as a raw material at another location. The $4.3 million facility to accomplish these results saves the company about $1 million a year, an attractive return on an investment that also eliminates the problem of disposal of a growing stream of sludge.

A relatively simple equipment change in any stage of a manufacturing process may promote waste reduction. When an employee discovered some gas escaping from a pressure control vent, a USX Chemicals plant in Ironton, Ohio, was able to reduce the amount emitted into the air by adding a condenser to existing equipment. The condensed gas was then returned directly to the phenol process unit. The company was able to recover 400,000 pounds of cumene, one of the plant's major raw materials. This procedure is saving the plant $200,000 in operating expenses annually. The installation cost of the condenser was only $5,000.

Operational changes include alterations in the way hazardous wastes are handled in a plant, such as spill minimization and more conservative use of chemicals. At its plant in Fremont, California, Borden was able to reduce phenol wastes by making a simple change in operational practices. Borden uses water to rinse the filters that remove large particles of phenol-containing resinous material from products as they are loaded into tank cars. Previously, this rinsewater was sent to floor drains where it contaminated all wastewater that was discharged to a municipal sewage treatment plant. With the change, the rinsewater is now collected in a 250-gallon recovery tank and reused as rinsewater when a new batch of phenolic resin is produced. The company is saving $50,000 a year in waste disposal costs and avoiding the legal expenses involved in applying for hazardous waste permits.

Simply by using a different cleaning solvent for the equipment used to produce vinyl benzal chloride, Dow Chemical Company was able to reduce hazardous waste generation by 70 percent.[37]

A timber company, through its research, developed a new use for tree bark, the last massive waste product of the wood-products industry. The firm designed a bark processor that made it the first domestic producer of vegetable wax, an important ingredient in cosmetics and polishes.

Incentives to do more along these lines could be provided in several ways. The producers could be subsidized to follow the desired approach. In this period of large budget deficits, that would, of course, increase the amount of money that the Treasury must borrow.

Another alternative is to tax the generation and disposal of wastes. The object would not be to punish the polluters but to get them to change their ways. If something becomes more expensive, business firms have a natural desire to use less of the item. In this case, the production of pollution would become more expensive. Every sensible firm would try to reduce the amount of pollution tax it pays by curbing its wastes. Adjusting to new taxes on pollution would be a matter not of altruism but of minimizing cost and maximizing profit. The pollution tax approach appeals to self-interest in order to achieve the public interest.

Charging polluters for the pollution they cause gives companies an incentive to find innovative ways to cut down on their discharges.[38] These fees would raise costs and hence prices for products whose production generates a lot of pollution. It is wrong to view this as a way of shifting the burden to the public. The relevant factor is that consumer purchasing is not static. Consumer demand would shift to products which pollute less, because they would cost less. To stay competitive, high-polluting producers would have to economize on pollution, just as they do in the case of other costs of production. Since pollution imposes burdens on the environment, it is only fair that the costs of cleanup be reflected in the price of a product whose production generates this pollution.

Nine countries in Western Europe have adopted the "polluter pays" principle. In these nations, pollution control is paid for directly by the polluting firm or from the money collected from effluent taxes. The West German effluent-fee system, the oldest in operation, began before World War I. It has succeeded in halting the decline in water quality throughout the Ruhr Valley, the center of West Germany's iron and steel production. It is also serving as a model for a more recent French effort.[39]

Indeed, as Table 7.1 shows, many foreign countries have shown a greater willingness than the United States to experiment with waste minimization approaches other than "command and control" measures. These approaches include: tax incentives, economic subsidies, technical assistance, demonstration projects, waste reduction and waste exchange programs.

Practical problems make changes in pollution policy difficult in the United States. Both the regulators and the regulated have an interest in maintaining the current approach. Pollution taxes have little appeal in the political system, particularly in the Congress. Many reject a pollution tax on philosophical grounds, considering pollution charges to be a "license to pollute." They believe that putting a price on the act of polluting amounts to an attitude of moral indifference towards polluters. That gets us back to the point made earlier, that many people look at ecological matters as moral issues, making it especially difficult to adopt a more rational and workable system.

Suggestions have been made for a more fundamental approach to cleaning up environmental wastes. Princeton University professors William J. Baumol and Edwin S. Mills conclude that the existing system focusing on "inspection and coercion" is useless, because there is no way of knowing who produces the wastes or who is currently storing wastes.

The Superfund experience, they contend, confirms that it is best to adopt measures that provide reasonable assurance that toxic wastes will be disposed of properly. In their view, only a financial incentive that succeeds in getting people to comply voluntarily has any chance of working. It also may prove less expensive than cleaning up the damage caused by indiscriminate dumping. Baumol and Mills would have the government identify a set of acceptable disposal facilities and pay for delivery of toxic wastes to the sites. The subsidy should not be set so high as to encourage deliberate production of hazardous waste—nor so low as to be ignored.[40]

TABLE 7.1
Methods of Encouraging Waste Minimization in Other Countries

Type	Japan	Canada	Germany	Sweden	Netherlands	Denmark
Tax Incentives						
Waste-End Taxes			x		x	x
Tax Incentives	x	x	x			
Economics						
Price Supports for Recycling			x		x	
Government Grants and Subsidies	x	x	x	x	x	x
Low-Interest Loans	x		x			
Technical Assistance						
Information Service	x	x	x		x	x
Site Consultation	x	x				
Training Seminars		x	x			x
R&D						
Technical Development Labs				x	x	x
Demonstration Projects	x	x	x	x		
Industrial Research			x	x	x	
Reduction Plans						
National Waste-Management Plans					x	
Waste-Reduction Agreements	x					
Waste-Reduction as a Part of Permits				x		
Waste Exchange						
Regional Waste Exchanges	x	x	x		x	x

Source: Environmental Protection Agency, *Waste Minimization*, 1987.

Yet, more technically efficient solutions will not stand a chance of enactment or even serious consideration in the legislative process until the "emotional thermostat" in public discussions on the environment is lowered substantially.

Conclusion

Although economists are often accused of being handmaidens of the business community, in practice environmental economics makes for strange alliances. So far, business interests have opposed the suggestions of economists for such sweeping changes in the basic structure of government regulation as using taxes or emissions fees to reduce pollution. Despite the shortcomings of the present system of government regulation, many firms have paid the price of complying with existing rules. They have learned to adjust to regulatory requirements and to integrate existing regulatory procedures into their long-term planning.

As any serious student of business-government relations will quickly report, the debate over regulation is miscast when it is described as black-hatted business versus white-hatted public interest groups. Almost every regulatory action creates winners and losers in the business system and often among other interest groups. Clean air legislation, focusing on ensuring that new facilities fully meet standards, is invariably supported by existing firms that are "grandfathered" approval without having to conform to the same high standards as new firms. Regulation thus protects the "ins" from the "outs."

There are many other examples of regulatory bias against change and especially against new products, new processes, and new facilities. Tough emissions standards are set for new automobiles, but not for older ones. Testing and licensing procedures for new chemicals are more rigorous and thoroughly enforced than for existing substances. This ability to profit from the differential impacts of regulation helps to explain why business shows little enthusiasm for the use of economic incentives and prefers current regulatory techniques.

But the reform of regulation is truly a consumer issue. The consumer receives the benefits from regulation and bears the burden of the costs of compliance in the form of higher prices and less product variety. Thus, the consumer has the key stake in developing more cost-effective solutions to environmental problems.

Notes

1. "Pollution in East Described as Grim," *The New York Times*, June 17, 1990, p. 7.

2. "Opinion Outlook," *National Journal*, April 28, 1990, p. 1052.

3. "Opinion Outlook," *National Journal*, October 27, 1990, p. 2619.

4. U.S. Comptroller General, *How to Dispose of Hazardous Waste* (Washington, D.C.: General Accounting Office, 1978), pp. 5-14.

5. Robert C. Mitchell and Richard T. Carson, "Protest, Property Rights, and Hazardous Waste," *Resources*, Fall 1986, p. 6.

6. Quoted in Geraldine Cox, *The Dangerous Myth of a Risk-Free Society*, an address to Drexel University, Philadelphia, Pa., February 19, 1987, p. 3.

7. *A Study of the Attitudes of the American Public Toward Improvement of the Natural Environment* (Washington, D.C.: National Wildlife Federation, 1969).

8. This section draws heavily from Murray L. Weidenbaum, *Business, Government and the Public*, Fourth Edition (Englewood Cliffs, N.J.: Prentice-Hall, 1990), Chapter 4.

9. Frank P. Grad et al., *Environmental Control* (New York: Columbia University Press, 1971), p. 49.

10. EPA Science Advisory Board, *Reducing Risk: Setting Priorities and Strategies for Environmental Protection* (Washington, D.C.: U.S. Environmental Protection Agency, 1990), p. 1.

11. Quoted in Peter Osterlund, "EPA at 15," *Christian Science Monitor*, December 5, 1985, p. 8.

12. Ibid.

13. *Environmental Fact Sheet: About the Municipal Solid Waste Stream* (Washington, D.C.: U.S. Environmental Protection Agency, August 1990).

14. U.S. Council on Environmental Quality, *Environmental Quality, 1990* (Washington, D.C.: U.S. Government Printing Office, 1990).

15. William D. Ruckelshaus, *Plateau of Hope: Some Perspectives on Environmental Achievement*, Remarks to the National Press Club, May 22, 1984, p. 11.

16. EPA Science Advisory Board, *Unfinished Business: A Comparative Assessment of Environmental Problems* (Washington, D.C.: U.S. Environmental Protection Agency, 1987); EPA Science Advisory Board, *Reducing Risk*, p. 3.

17. National Research Council, Committee on Scientific and Regulatory Issues Underlying Pesticide Use Patterns and Agricultural Innovation, *Regulating Pesticides in Food* (Washington, D.C.: National Academy Press, 1987).

18. U.S. Comptroller General, *Hazardous Waste: Federal Civil Agencies Slow to Comply with Regulatory Requirements* (Washington, D.C.: General Accounting Office, 1986).

19. U.S. Comptroller General, *Superfund: Civilian Federal Agencies Slow to Clean Up Hazardous Waste* (Washington, D.C.: General Accounting Office, 1987), p. 3.

20. Vic Fazio, "Needed: A Military Superfund," *Christian Science Monitor*, October 3, 1985, p. 18.

21. U.S. Comptroller General, *Hazardous Waste Management at Tinker Air Force Base* (Washington, D.C.: General Accounting Office, 1985), p. 4.

22. See, for example, Andy Bigford, "City Will Stop River Dumping," *Aspen Daily News*, July 21, 1987, p. 1.

23. *Dioxin in the Environment* (Summit, N.J.: American Council on Science and Health, 1984), p. 5.

24. U.S. Comptroller General, *Illegal Disposal of Hazardous Waste* (Washington, D.C.: General Accounting Office, 1985), p. i.

25. "Waste Not, Want Not," *Journal of American Insurance*, Third Quarter 1987, p. 3.

26. Mitchell and Carson, "Hazardous Waste," pp. 6-8.

27. My colleague James Davis suggests a more focused subsidy—the payment of cash to each resident of the area with the payment rising, or falling, with the distance from the dump site.

28. Comptroller General, *Hazardous Waste*, 1986, p. 9.

29. "Cancer and Environment: Higginson Speaks Out," *Science*, September 28, 1979, p. 1363.

30. Harry Demopoulos, *Environmentally Induced Cancer . . . Separating Truth From Myth*, a talk to the Synthetic Organic Chemical Manufacturers Association, Hasbrouck Heights, New Jersey, October 4, 1979, pp. 1-6. For another study with similar conclusions, see R. Doll and R. Peto, "The Causes of Cancer: Quantitative Estimates of Avoidable Risks of Cancer in the United States Today," *Journal of the National Cancer Institute*, Vol. 66, No. 6, 1981, pp. 1191-1308.

31. U.S. Bureau of the Census, *Statistical Abstract of the United States, 1987* (Washington, DC: U.S. Government Printing Office, 1986), p. 69.

32. Bruce N. Ames, *Six Common Errors Relating to Environmental Pollution* (Louisville, Ky.: National Council for Environmental Balance, 1987).

33. U.S. Office of Technology Assessment, *Annual Report to the Congress, Fiscal Year 1986* (Washington, D.C.: U.S. Government Printing Office, 1986), p. 41.

34. Joel S. Hirschhorn and Kirsten U. Oldenburg, "Preventing Pollution Is No End-of-Pipe Dream," *Across the Board*, June 1987, p. 11.

35. "Superfund—It's Time for a Waste-End Fee," *CMA News*, January 1984, p. 16.

36. "Case Histories Detailed—Corporate Imagination, Cooperation, Practicality Lead to Environmental Progress," *Roundtable Report*, March 1984, pp. 1-3.

37. Joanna Underwood, "Eliminate Toxic Wastes and Increase Your Profits," *World Link*, April 1988, p. 25.

38. Allen Kneese et al., eds., *Managing the Environment* (New York:

Praeger, 1971); Frederick R. Anderson et al., *Environmental Improvement Through Economic Incentives* (Baltimore: Johns Hopkins University Press, 1978).

39. Craig E. Reese, *Deregulation and Environmental Quality* (Westport, Conn.: Quorum Books, 1983), pp. 95-105, 147-153.

40. William J. Baumol and Edwin S. Mills, "Paying Companies to Obey the Law," *The New York Times*, October 27, 1985, p. F3.

Bibliography

Alliance Technologies Corporation. *Cost Assessment of Alternative National Ambient Air Quality Standards for Ozone.* Draft report, prepared for the U.S. EPA Office of Air Quality Planning and Standards, October 1987.

American Cancer Society. *Cancer Facts & Figures—1990.* Atlanta: ACS, 1990.

American Council on Science and Health. *Dioxin in the Environment.* Summit, N.J.: ACSH, 1984.

American Petroleum Institute. *Ozone Concentration Data.* Washington, D.C.: API, March 1989.

Ames, Bruce N. "Dietary Carcinogens and Anticarcinogens." *Science* 221 (1983): 1256-64.

———. Reply to "Letters." *Science* 224 (1984): 760.

———. "Water Pollution, Pesticide Residues, and Cancer," adapted from November 11, 1985, testimony to the Senate Committee on Toxics and Public Safety Management. *Water* 27/2 (1985/1986): 23-24.

———. *Six Common Errors Relating to Environmental Pollution.* Louisville, Ky.: National Council for Environmental Balance, 1987.

Ames, Bruce N., R. Magaw, and L. S. Gold. "Ranking Possible Carcinogenic Hazards." *Science* 236 (April 17, 1987): 271-80.

Anderson, Elizabeth L. "The Risk Analysis Process," in C. C. Travis, ed., *Carcinogen Risk Assessment*, pp. 3-17. New York: Plenum Press, 1988.

Anderson, Frederick R., et al. *Environmental Improvement Through Economic Incentives.* Baltimore: Johns Hopkins University Press, 1978.

Balling, R. C., and R. S. Cerveney. Article in *Journal of Applied Meteorology* 27 (1988): 881.

Bottomly, M., et al. *Global Ocean Surface Temperature Atlas.* Cambridge: MIT Department of Earth and Planetary Science, 1990.

Bryan, K., et al. Article in *Science* 215 (1982): 56.

"Buried Alive." *Newsweek* (November 27, 1989): 70.

Byrd, Daniel, and Lester B. Lave. "Narrowing the Range: A Framework for Risk Regulators." *Issues in Science and Technology* 3 (Summer 1987): 92-97.

Carson, Rachel. *Silent Spring*. Greenwich, Conn.: Fawcett, 1962.

"Case Histories Detailed—Corporate Imagination, Cooperation, Practicality Lead to Environmental Progress." *Roundtable Report* (March 1984): 1-3.

Chilton, Kenneth, and Anne Sholtz. *Battling Smog: A Plan for Action*, Formal Publication No. 93. St. Louis: Center for the Study of American Business, 1989.

Cicerone, R. J. "Methane in the Atmosphere," in S. F. Singer, ed., *Global Climate Change: Natural and Human Influences*. New York: Paragon House Publishers, 1989.

Commoner, Barry. *The Closing Circle: Nature, Man and Technology*. New York: Bantam Press, 1972.

———. *Making Peace with the Planet*. New York: Pantheon Books, 1990.

Cox, Geraldine. *The Dangerous Myth of a Risk-Free Society*, an address to Drexel University. Philadelphia, Pa.: February 19, 1987.

d'Arge, R., et al. From *CIAP Monograph 6*. DOT-TST-75-56. Washington, D.C.: U.S. Department of Transportation, 1975.

Davis, Devra L., and B. H. Magee. "Cancer and Industrial Chemical Production." *Science* 206 (December 21, 1979): 1356.

Demopoulos, Harry. *Environmentally Induced Cancer . . . Separating Truth From Myth*, a talk to the Synthetic Organic Chemical Manufacturers Association, pp. 1-6. Hasbrouck Heights, N. J.: October 4, 1979.

Doll, R., and R. Peto. *The Causes of Cancer*. Oxford: Oxford University Press, 1981.

———. "The Causes of Cancer: Quantitative Estimates of Avoidable Risks of Cancer in the United States Today." *Journal of the National Cancer Institute* 66, no. 6 (1981): 1191-1308.

Echard, Jo Kwong. *Protecting the Environment: Old Rhetoric, New Imperatives*. Washington, D.C.: Capital Research Center, 1990.

Efron, Edith. *The Apocalyptics: Cancer and the Big Lie*. New York: Simon and Schuster, 1984.

Ellsaesser, H. W. Article in *Atmospheric Environment* 18 (1984): 431.

———. Paper in *Symposium on Global Climate Change*, proceedings from a conference held July 30-31, 1990. Los Angeles: California Energy Commission, 1990.

Fact Sheet: President Bush's Clean Air Plan. Washington, D.C.: The White House, June 12, 1989.

Flavin, C. "Slowing Global Warming." *State of the World*. Washington, D.C.: Worldwatch Institute, 1990.

"From Microbes to Men: The New Toxic Substances Control Act and Bacterial Mutagenicity/Carcinogenicity Tests." *Environmental Law Reporter* 6 (1976): 10251-252.

Glaser, P. E. "Power from the Sun: Its Future." *Science* 162 (1968): 857.

Goldman, Marshall. *The Spoils of Progress: Environmental Pollution in the Soviet Union*. Cambridge: MIT Press, 1972.

Goode, Stephen. "Artists Draw Line at Cadmium." *Insight* (July 16, 1990): 60.

Grad, Frank P., et al. *Environmental Control.* New York: Columbia University Press, 1971.

Graham, J. D., L. C. Green, and M. J. Roberts. *In Search of Safety: Chemicals and Cancer Risks.* Cambridge: Harvard University Press, 1988.

Handler, Philip. "Dedication Address," Northwestern University Cancer Center, 1979.

Hardin, Garrett. "The Tragedy of the Commons." *Science* (December 13, 1968): 1244-45.

Higginson, John. "Cancer and Environment: Higginson Speaks Out," reported by T. H. Maugh. *Science* 205 (September 28, 1979): 1363-66.

Hirschhorn, Joel S., and Kirsten U. Oldenburg. "Preventing Pollution Is No End-of-Pipe Dream." *Across the Board* (June 1987): 11.

Horwitz, Morton J. *The Transformation of American Law, 1780-1860.* Cambridge: Harvard University Press, 1977.

Houghton, J. T., G. J. Jenkins, and J. J. Ephraums, eds. *Climate Change: The IPCC Scientific Assessment.* Cambridge: Cambridge University Press, 1990.

Hutt, P. B. "Use of Qualitative Risk Assessment in Regulatory Decisionmaking Under Federal Health and Safety Statutes," in D. G. Hoel, R. A. Merrill, and F. P. Perera, eds., *Risk Quantitation and Regulatory Policy.* Cold Spring Harbor, N.Y.: Cold Spring Harbor Laboratory, 1985.

Idso, S. B. *Carbon Dioxide and Global Change: Earth in Transition.* Tempe, Ariz.: IBR Press, 1989.

Jones, P. D., T. M. L. Wigley, and P. B. Wright. Article in *Nature* 322:430.

Jones, Troyce. "Radiological and Chemical Contamination: Should I Spend Money for the Present or the Future?" *The Health Physics Society Newsletter* XV/5 (1987): 1-4.

Karl, T. R., et al. Article in *Journal of Climate and Applied Meteorology* 23 (1984): 1489.

———. *Historical Climatology Series 4-5.* Asheville, N.C.: NOAA National Climate Data Center, 1988.

King, Llewelyn. "Nuclear Power in Crisis: The New Class Assault." *Energy Daily* 5/135 (1978): 7.

Kneese, Allen, et al., eds. *Managing the Environment.* New York: Praeger, 1971.

Kriz, Margaret E. "It's Not Easy Buying Green." *National Journal* (July 14, 1990): 1745.

Krupnick, Alan J., and Raymond J. Kopp. *The Health and Agricultural Benefits of Reductions in Ambient Ozone in the United States.* Washington, D.C.: Resources for the Future, Discussion Paper QE88-10, August 1988.

Lange, Oscar, and Fred M. Taylor. *On the Economic Theory of Socialism.* Minneapolis: University of Minnesota Press, 1938.

Lave, Lester B. *The Strategy of Social Regulation.* Washington, D.C.: Brookings Institution, 1981.

———. "Health and Safety Analysis: Information for Better Decisions." *Science* 236 (April 17, 1987): 291-95.

Lave, Lester B., F. K. Ennever, H. S. Rosenkranz, and G. S. Omenn. "Information Value of the Rodent Bioassay." *Nature* 336 (December 15, 1988): 631-33.

Lindzen, R. J. Article in *Bulletin of the American Meteorological Society* 71 (1990): 288.

Manne, A. S., and R. G. Richels. *Global CO2 Emissions Reductions: The Impacts of Rising Energy Costs.* Palo Alto: Stanford University and Electric Power Research Institute, February 1990.

Martin, J., et al. Article in *Nature* 345 (1990): 516.

Matlack, Carol. "Recycling Bandwagon." *National Journal* (September 30, 1989): 2399-2402.

McDonnell, W. F., D. H. Horstman, S. Abdul-Salaam, and D. E. House. "Reproducibility of Individual Responses to Ozone Exposure." *American Review of Respiratory Disease* 131 (1985): 36.

McKee, Herbert. "Valid Measurements: Key to Decisions." *Chemtech* 25 (1985): 431-33.

Milvy, Paul. "Actual and Perceived Risks from Chemical Carcinogens." *Risk Analysis* 6 (March 1986): 69-80.

Mitchell, Robert C., and Richard T. Carson. "Protest, Property Rights, and Hazardous Waste." *Resources* (Fall 1986): 6.

National Research Council. *Toxicity Testing.* Washington, D.C.: National Academy Press, 1984.

———. Committee on Scientific and Regulatory Issues Underlying Pesticide Use Patterns and Agricultural Innovation. *Regulating Pesticides in Food.* Washington, D.C.: National Academy Press, 1987.

National Solid Wastes Management Association. *Landfill Capacity in the Year 2000.* Washington, D.C.: NSWMA, 1989.

———. *Public Attitudes Toward Garbage Disposal.* Washington, D.C.: NSWMA, 1990.

National Wildlife Federation. *A Study of the Attitudes of the American Public Toward Improvement of the Natural Environment.* Washington, D.C.: NWF, 1969.

Newell, R. E., et al. Article in *Geophysical Research Letters* 16 (1989): 32.

Nordhaus, W. D. Article in H. Aaron, ed., *Setting National Priorities: Policy for the Nineties.* Washington, D.C.: Brookings Institution, 1990, pp. 185-211.

Pigou, Arthur C. *Wealth and Welfare.* London: Macmillan, 1912.

Rathje, William. "Rubbish." *Atlantic Monthly* (December 1989): 103.

Reese, Craig E. *Deregulation and Environmental Quality.* Westport, Conn.: Quorum Books, 1983.

Reifsnyder, W. E. Article in *Agricultural and Forest Meteorology* 47 (1989): 349.

Rogers, L. B., cited in "Trace Analysis: Bringing Certainty to Confu-

sion." *The Point Is . . .* (public issues newsletter from Dow Chemical Company) 47 (1981): 2.
Ross, Malcolm. "A Definition for Asbestos," in *ASTM Special Technical Publication 834*, pp. 139-47. Philadelphia: ASTM, 1984.
———. "A Survey of Asbestos-Related Diseases in Trades and Mining Occupations and in Factory and Mining Communities as a Means of Predicting Health Risks of Nonoccupational Exposure to Fibrous Minerals," in *ASTM Special Technical Publication 834*, pp. 51-104. Philadelphia: ASTM, 1984.
Rowland, F. S. "Chlorofluorocarbons, Stratospheric Ozone, and the Antarctic Ozone Hole," in S. F. Singer, ed., *Global Climate Change: Natural and Human Influences.* New York: Paragon House Publishers, 1989.
Ruckelshaus, William D. *Plateau of Hope: Some Perspectives on Environmental Achievement.* Remarks to the National Press Club, May 22, 1984.
Russell, Milton. *Ozone Pollution: The Hard Choices.* Oak Ridge, Tenn.: Oak Ridge National Laboratory, 1988.
Saffiotti, Umberto. "Risk-Benefit Considerations in Public Policy on Environmental Carcinogenesis," in *Proceedings of the Eleventh Canadian Cancer Research Conference.* Toronto: National Cancer Institute, 1976.
Schmidt, Alexander. Quoted by T. H. Jukes, "DES in Beef Production: Science, Politics and Emotion," in H. H. Hiatt, J. D. Watson, and J. A. Winsten, eds., *Origins of Human Cancer*, pp. 1658-59. Cold Spring Harbor, N.Y.: Cold Spring Harbor Laboratory, 1977.
Schneider, S. M. "Climate Modeling." *Scientific American* 256 (May 1987): 72-80.
Sharp, D. S., B. Eskenazi, R. Harrison, P. Callas, and A. H. Smith. "Delayed Health Hazards of Pesticide Exposure," in L. Breslow, J. Fielding, and L. Lave, eds., *Annual Review of Public Health*, pp. 441-71. Palo Alto, Calif.: Annual Reviews, 1986.
Sholtz, Anne. *The Cost-Effectiveness of Proposed Measures for Controlling Refueling Emissions*, Working Paper No. 131. St. Louis: Center for the Study of American Business, 1989.
Singer, S. F. "The Responses to Climate Change: The Significance of Different Time Scales." *Living with Climate Change.* McLean, Va.: MITRE Corp., 1977.
———. "Postscript," in Singer, ed., *Global Climate Change: Natural and Human Influences.* New York: Paragon House Publishers, 1989.
Spencer, R. W., and J. R. Christy. Article in *Science* 247 (1990): 1558.
"Superfund—It's Time for a Waste-End Fee." *CMA News* (January 1984): 16.
Teller, Azriel. "Air Pollution Abatement: Economic Rationality and Reality," in Roger Reville and Hans Landsberg, eds., *America's Changing Environment.* Boston: Beacon Press, 1970.
Timberlake, Lloyd. "Poland—The Most Polluted Country in the World?" *New Scientist* (October 22, 1981): 249.

Totter, John R. "Spontaneous Cancer and Its Possible Relationship to Oxygen Metabolism." *Proceedings of the National Academy of Sciences*, 1980.

Travis, C. C., ed. *Carcinogen Risk Assessment*. New York: Plenum Press, 1988.

U.S. Bureau of the Census. *Statistical Abstract of the United States, 1987*. Washington, D.C.: Government Printing Office, 1986.

U.S. Comptroller General. *How to Dispose of Hazardous Waste*. Washington, D.C.: General Accounting Office, 1978.

———. *Hazardous Waste Management at Tinker Air Force Base*. Washington, D.C.: General Accounting Office, 1985.

———. *Illegal Disposal of Hazardous Waste*. Washington, D.C.: General Accounting Office, 1985.

———. *Hazardous Waste: Federal Civil Agencies Slow to Comply with Regulatory Requirements*. Washington, D.C.: General Accounting Office, 1986.

———. *Superfund: Civilian Federal Agencies Slow to Clean Up Hazardous Waste*. Washington, D.C.: General Accounting Office, 1987.

———. *Wastepaper Recycling: Programs of Civil Agencies Waned During the 1980s*. Washington, D.C.: General Accounting Office, 1989.

U.S. Council on Environmental Quality. *Environmental Quality, 1990*. Washington, D.C.: Government Printing Office, 1990.

U.S. Department of Health, Education, and Welfare. "Estimates of the Fraction of Cancer in the United States Related to Occupational Factors," reprinted in R. Peto and M. Schneiderman, eds., *Quantification of Occupational Cancer*. Cold Spring Harbor, N.Y.: Cold Spring Harbor Laboratory, 1981.

U.S. Environmental Protection Agency. *Air Quality Criteria for Ozone and Other Photochemical Oxidants*, Vols. I and V. Washington, D.C., August 1986.

———. *Waste Minimization*. Washington, D.C., 1987.

———. *Draft Regulatory Impact Analysis: Control of Gasoline Volatility and Evaporative Hydrocarbon Emissions from New Motor Vehicles*, Tables 3-21, 3-22, 6-3, 6-4, 6-5, 6-6. Washington, D.C., July 1987.

———. "EPA Lists Places Failing to Meet Ozone or Carbon Monoxide Standards." *Environmental News* (July 27, 1987): Table 1.

———. *Review of the National Ambient Air Quality Standards for Ozone and Other Photochemical Oxidants*. Washington, D.C.: Office of Air Quality Planning and Standards, November 1987.

———. *National Air Quality and Emissions Trends Reports, 1987*. Washington, D.C., March 1989.

———. *National Emissions Standards for Hazardous Air Pollutants, Benzene Emissions*. Washington, D.C., August 31, 1989.

———. *Environmental Fact Sheet: About the Municipal Solid Waste Stream*. Washington, D.C., August 1990.

———. Science Advisory Board. *Unfinished Business: A Comparative Assessment of Environmental Problems.* Washington, D.C., 1987.

———. *Reducing Risk: Setting Priorities and Strategies for Environmental Protection.* Washington, D.C., 1990.

U.S. General Accounting Office. *Air Pollution: Ozone Attainment Requires Long Term Solutions to Solve Complex Problems.* Washington, D.C., January 1988.

U.S. National Cancer Institute. *Cancer Control Objectives for the Nation: 1985-2000.* Bethesda, Md.: USNCI, 1986.

U.S. Office of Technology Assessment. *Annual Report to the Congress, Fiscal Year 1986.* Washington, D.C.: U.S. Government Printing Office, 1986.

———. *Facing America's Trash: What Next for Municipal Solid Waste?* Washington, D.C., 1989.

———. *Catching Our Breath: Next Steps for Reducing Urban Ozone.* Washington, D.C., July 1989.

Underwood, Joanna. "Eliminate Toxic Wastes and Increase Your Profits." *World Link* (April 1988): 25.

Washington, W. M., and G. A. Meehl. Article in *Climate Dynamics* 4 (1989): 1.

"Waste Not, Want Not." *Journal of American Insurance*, Third Quarter, 1987: 3.

Weidenbaum, Murray L. *Business, Government and the Public*, Fourth Edition. Englewood Cliffs, N.J.: Prentice-Hall, 1990.

Weill, Hans, and Janet M. Hughes. "Asbestos as a Public Health Risk: Disease and Policy," in L. Breslow, J. Fielding, and L. Lave, eds., *Annual Review of Public Health*, pp. 171-92. Palo Alto, Calif.: Annual Reviews, 1986.

Whalen, Elizabeth M. *Toxic Terror.* Ottawa, Ill.: Jameson Books, 1985.

Whipple, Chris, ed. *De Minimis Risk: Contemporary Issues in Risk Analysis*, Vol. 2. New York: Plenum Press, 1987.

Wildavsky, Aaron. "No Risk Is the Highest Risk of All." *American Scientist* 67, no. 1 (January-February 1979): 32-37.

———. "Richer Is Safer." *The Public Interest* 60 (1980): 23-29.

Wildavsky, Aaron, and Mary Douglas. *Risk and Culture.* Berkeley: University of California Press, 1982.

Winner, Langdon. *Autonomous Technology.* Cambridge: MIT Press, 1977.

Zwally, H. J. Article in *Science* 246 (1989): 1589.

About the Editors and Contributors

Kenneth Chilton is the associate director of the Center for the Study of American Business at Washington University in St. Louis.

Thomas DiLorenzo is Scott L. Probasco, Jr., Professor of Free Enterprise and Director of the Center for Economic Education at the University of Tennessee at Chattanooga.

Lester Lave is the James H. Higgins Professor of Economics at Carnegie-Mellon University in Pittsburgh.

Margaret Maxey is Professor of Bioethics and Director of the Clint W. Murchison, Sr., Chair of Free Enterprise in the College of Engineering at the University of Texas at Austin.

S. Fred Singer, on leave from the University of Virginia, is director of the Science and Environmental Policy Project at the Washington Institute for Values in Public Policy.

Anne Sholtz-Vogt is a doctoral candidate in economics at Washington University in St. Louis.

Melinda Warren is writer/analyst of the Center for the Study of American Business at Washington University in St. Louis.

Murray Weidenbaum is director of the Center for the Study of American Business and Mallinckrodt Distinguished University Professor at Washington University in St. Louis.

Index